Environmental Technology Development in Liberal and Coordinated Market Economies

Environmental Technology Development in Liberal and Coordinated Market Economies

Tweaking Institutions

Shiu-Fai Wong

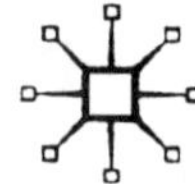

ENVIRONMENTAL TECHNOLOGY DEVELOPMENT IN LIBERAL AND COORDINATED MARKET ECONOMIES
© Shiu-Fai Wong, 2006.

First published in 2006 by
PALGRAVE MACMILLAN™
175 Fifth Avenue, New York, N.Y. 10010 and
Houndmills, Basingstoke, Hampshire, England RG21 6XS
Companies and representatives throughout the world.

PALGRAVE MACMILLAN is the global academic imprint of the Palgrave Macmillan division of St. Martin's Press, LLC and of Palgrave Macmillan Ltd. Macmillan® is a registered trademark in the United States, United Kingdom and other countries. Palgrave is a registered trademark in the European Union and other countries.

ISBN-13: 978–1–4039–7642–0
ISBN-10: 1–4039–7642–2

Library of Congress Cataloging-in-Publication Data is available from the Library of Congress.

A catalogue record for this book is available from the British Library.

Design by Newgen Imaging Systems (P) Ltd., Chennai, India.

First edition: December 2006

10 9 8 7 6 5 4 3 2 1

Printed in the United States of America.

*For those who are interested in improving
state governance and our planet.*

Contents

List of Charts and Tables

Charts

Tables

List of Abbreviations

AAQS	Ambient Air Quality Standards
ACE	Advanced Clean Energy (Vehicles Projects)
ACOR	Australia Council of Recyclers
ACP	Advisory Committee on Packaging
ANZECC	Australian and New Zealand Environment and Conservation Council
ARPWO	Avoidance and Recovery of Packaging waste Ordinance
AT-PZEV	Advanced Technology Partial Zero Emission Vehicles
BIEC	Beverage Industry Environment Council
BMBF	*Bundesministerium für Bildung und Forschung* (the existing name of the Federal Ministry for Research and Technology)
BMFT	*Bundesministerium für Forschung und Technologie* (the former name of the Federal Ministry for Research and Technology)
BMWA	*Bundesministerium für Wirtschaft und Arbeit* (the existing name of the Ministry of Economics and Technology)
BWE	*Bundesverband Windenergie eV*
BWEA	British Wind Energy Association
BWMi	*Bundesminister für Wirtschaft und Technologie* (the former name of the Ministry of Economics and Technology)
CAFE	Corporate Average Fuel Economy
CARB	Californian Air Resources Board
CAT	Columbians Against Throwaways
CCL	Climate Change Levy
CDL	Container Deposit Legislation
CDU	*Christlichdemocratische Union Deutschland* (Christian Democratic Union)
CHP	Combined Heat and Power (Policy)

CLEAR	Clean Efficient Autos Resulting (from Advanced Car Technologies)
CME	Coordinated Market Economy
CNG	Compressed Natural Gas
CRS	Curbside Recycling Systems
CSU	*Christlichsoziale Union* (Christian Social Union)
D-CT	Democrat-Connecticut
DEFRA	Department for Environment Food and Rural Affairs
DETR	Department of the Environment, Transport and the Regions
DfT	Department for Transport
D-HI	Democrat-Hawaii
D-MA	Democrat-Massachusetts
DOD	Department of Defense
DOT	Department of Transport
DSD	Duales System Deutschland (AG)
DTI	Department Trade and Industry
DTLR	Department for Transport, Local Government and the Regions
D-VT	Democrat-Vermont
EA	Embedded Autonomy (Concept)
EA	Environment Agency
EC	European Community
ECN	Energy Research Centre of the Netherlands
EEC	European Economic Community
EEG	*Erneuerbare-Energien-Gesetz* (Renewable Energy Law)
EFA	Environmentally Friendly Auto
EFL	Electricity Feed Law
EFP	*Energieforschungsprogramm* (Energy Research Program)
EPA	Environmental Protection Agency
ERT	Energy Research and Technology
E-T	Environmental-Technology
ETSU	Energy Technology Support Unit
EU	European Union
EWEA	European Wind Energy Association
FCCSET	Federal Council for Science, Engineering and Technology
FDP	Freie Democratische Partei (Free Democratic Party)
FFL	Fossil Fuel Levy
GI	Governed Interdependence (Concept)
GM	Governed Market (Concept)
GWEA	German Wind Energy Association
ICE	Internal Combustion Engine

IEA	International Energy Agency
IEAHEV	International Energy Agency Hybrid and Electric Vehicle
IPP	Independent Power Producer
IRA	International Regime Approach
ITS	Intelligent Transport System
I-VA	Independent-Virginia
JETRO	Japan External Trade Organization
LME	Liberal Market Economy
LPA	Local Planning Authority
LPG	Liquefied Petroleum Gas
MAFF	Ministry of Agriculture, Fisheries and Food
METI	Ministry of Economy, Trade and Industry
MITI	Ministry of International Trade and Industry
MPG	Miles Per Gallon
MTBE	Methyl Tertiary-Butyl Ether
MW	Mega Watt
NAAQS	National Ambient Air Quality Standards
NASA	National Aeronautics and Space Administration
NEPM	National Environment Protection Measure (on Used Packaging)
NFFO	Non-Fossil Fuel Obligation
NFPA	Non-Fossil Purchasing Agency
NIC	Newly Industrializing Country
NI-NFFO	Northern Ireland Non-Fossil Fuel Obligation
NPA	National Patterns Approach
NPC	National Packaging Covenant
NRC	National Research Council
NSF	National Science Foundation
OECD	Organisation for Economic Co-Operation and Development
OFFER	Office of Electricity Regulation
OFGAS	Office of Gas Supply
OFGEM	Office of Gas and Electricity Markets
OPRP	Ordinance on Producers' Responsibility for Packaging
PET	Polyethylene Terephthalate
PNGV	Partnership for a New Generation of Vehicles
PP	Pacifying Permeation (Concept)
PRN	Packaging Waste Recovery Note
PRO	Producer Responsibility Obligation
PRS	PET-Recycling Schweiz
PSA	Policy Sector Approach

PWR	Packaging Waste Recycling
PZEV	Partial Zero-Emission Vehicle
RD&D	Research, Development and Demonstration
RES	Renewable Energy Systems
RPS	PET-Recycling Schweiz
RVM	Reverse Vending machines
SEPA	Scottish Environment Protection Agency
SFS	*Sveriges Förenade Studentkårer* (Swedish Arbitration Act)
SIP	State Implementation Plan
SPD	*Sozialdemokratische Partei Deutschlands* (Social Democrat Party)
SRO	Scottish Renewables Obligation
SULEV	"Super-Ultra-Low" Emission Vehicle
TLP	Targeted Leading Product (Program)
TPA	Policy Sector Approach
USABC	U.S. Advanced Battery Consortium
UVE	*Udviklingsprogram for Vedvarende Energi* (Program for Development, Demonstration and Information of Renewable Energy)
VER	Voluntary Export Restraint
WECS	Wind Energy Conversion System
WMEP	*Wissenschaftliches Meß und Evaluierungsprogramm* (Scientific Measurement and Evaluation Program)
ZEV	Zero-Emission Vehicle

PREFACE

The reasons behind the failure of many environmental technologies (E-Ts) in advanced industrial countries have puzzled scholars and policymakers for decades. Even traditional government policies such as the implementation of market-oriented developmental programs, "pick-the-winner" industrial policies (or through the delegation of powers to industry), and the encouragement of public-private partnerships have never really worked as per plan. Do institutions really matter? Would any other policies render renewable energy, waste recycling, clean fuel auto technology, and the like more indispensable to our daily life than the traditional ones? Existing institutional comparisons and concepts such as Hall and Soskice's institutional comparative advantage suggest that different institutional complementarities will have different impacts on national development. By comparing E-T development processes in different institutional structures, this book finds that the recent success of institutional reform in some E-T sectors have not only shared the view of such institutional comparative advantage but have also revealed the inevitability of the subtle use of state power and regulation. While many scholars are keen to highlight the advantages and disadvantages of liberal market economies and coordinated market economies, few have studied how governments have been able to transform their own institutional comparative disadvantages into advantages. Under the relentless pressure of environmental degradation and resource scarcity, advanced industrial countries have had little choice but to boldly and carefully devise and implement institutions capable of faster and more effective environmental-technology diffusion. Many scholars have explored state-societal structures and relationships as possible sources of institutional competitive advantage. However, the serious issue of conflict of interest between government, industry-dominated polluters, and E-T developers has never been touched on in previous institutional studies. This gap in knowledge is the theme of this book. It is argued here that institutions are fundamental to E-T development but are subject to new contingencies in place of older ones. This new institutional status quo is commonly found in all of the three sampled E-T sectors and all of the seven sampled countries with divergent political economies, demonstrating how the insufficiency of market forces and "the rule of law" can be supplemented and tweaked by "the rule of bureaucracy," especially through a new form of interdependent relationship between state and society, in dealing with the unique problems of the E-T sector.

Acknowledgments

I am grateful to many people who have contributed to the completion of this book. Without them, the accomplishment of this academic requirement could not have been possible. I first thank Professor Linda Weiss for her tremendously useful advice and comments on my research work. I thank other academics in the Discipline of Government and International Relations, especially Professor Graeme Gill, Dr. John M. Hobson, and Dr. Darryl Jarvis, for their inspiring comments in my seminars.

Special thanks are due to Professor Staffan Jacobsson from the Chalmers University of Technology, Dr. M. Ramesh from the National University of Singapore, and Dr. Rachel Parker from the University of Queensland for examining the entire manuscript and giving me many useful comments.

I thank people from other fields who read and edited my manuscripts: Stewart Packham, and especially Sung-Young Kim who did an exhaustive editing of the final version of the book at an incredible speed. Sincere thanks also go to my research fellows, particularly Dr. Susan Park, Shelly Savage, Reg Fisk, Lian-Siew Han, Young-Ju Kwon, Krisztina Molnar, Takashi Tsukamoto, Ladan Rahmani, Soo-Jin Yoon, John Mikler, John Warburton, Kathy Kang, Andrew Jenkinson, Dwane Byrne, James Young, and Marcus Chadwick, for all their support. I also want to thank my colleagues and friends from other departments in the Economics and Business Building, especially Dr. Tony Stapledon and Kozo Otsuka, for their continued encouragement.

I am similarly most grateful to the administrative staff, particularly Lynelle Rodrigues and Maria Robertson, for maintaining a pleasant and conducive environment and a well-equipped carrel to work in and for their efficient coordination during my research study.

Professor Martin Painter, the principle researcher of my other projects, in the Department of Public and Social Administration at the City University of Hong Kong has helped to advance my thoughts so that I could revise the manuscript more smoothly. His support is thus gratefully acknowledged.

My former bosom friends and colleagues, Christopher Wong, deputy general manager at Siemens Guangzhou, Hans Klug and Rolf Sallmüller, senior managers at MAN Augsberg, Thomas Holzinger, head of financing at MAN Munich, and many other helpful colleagues at MAN B&W Hong Kong and Siemens Erlangen provided accurate and detailed ideas of how infrastructural projects and electricity markets are run in different countries. With their many regular intelligent inputs, they helped me form the backbone of this book.

My thanks are also extended to Anthony Wahl and Heather VanDusen at Palgrave Macmillan. They deserve special mention not only for their efforts in studying and recommending the book but also for the excitement and enthusiasm they showed for the project. In addition, I give special thanks to Maran Elancheran, the book editor who handled the packaging of the book and supervised most of the production work.

What is more, years of personal hardships could have not been overcome without the love, prayers, and resource support of my family in Hong Kong. The support helped me stay focused at work. This endeavor in Sydney is dedicated to them.

About the Author

Shiu-Fai Wong holds a Ph.D. in Economics from the University of Sydney. He has a decade-long experience in infrastructural projects working with German-Danish consortia and East Asian governments. His research work in the comparative and international politics relates to the policy sectors of environmental technology, telecommunications industry, and capital market. He is currently a senior research associate at the City University of Hong Kong. His recent publications include "Obliging Institutions and Industry Evolution: A Comparative Study of the German and U.K. Wind Energy Industries," *Industry and Innovation*, Vol. 12, No. 1 (March 2005) and two more recent journal papers titled "Varieties of the Regulatory State: Government-Business Relations and Telecommunications Reforms in Malaysia and Thailand" and "The Telecommunications Regulatory Regimes in Hong Kong and Singapore: When Direct State Intervention Meets Indirect Policy Instruments," both coauthored with Professor Martin Painter in 2006.

1

Traditional Institutions are Ineffective in an Exceptional Policy Sector

Introduction

In response to the two oil crises and mounting concerns about environmental degradation, many western countries during the 1970s began to reconsider their energy policies and started developing environmental-technology (E-T) programs that primarily involved government-funded research and demonstration projects targeting a variety of E-Ts. Creating a wide range of technological tools—pipe-end oxidation catalyzers, solar cells, wind generators, clean-fuel cars, recycling appliances, waste incinerators, and so on—these programs have rapidly expanded over the past generation. Yet, there are plenty of E-T development problems such as poor commercialization performance and low diffusion rates. Studies have commonly attributed these problems to several factors, including poor technological productivity, severe market and systemic failure,[1] inadequate infrastructure, unfair politicization and externality, and/or insufficient local participation.

Whatever the specific problems behind the scene, it is widely agreed that governments are the ones who can do something to improve these factors, even though there might also be some tradeoffs (e.g., Jaffe, Newell, and Stavins 1999; Snyder, Millers, and Stavins 2003; Stavins 2004). Therefore, governments organize research, development, and demonstration (RD&D) programs to improve E-T productivity; establish laws, regulations and technological/organizational linkages to fix market and systemic failure; build public facilities to provide sufficient infrastructure; install technocracies, exclusive agencies, and/or independent regulators to reduce unfair political decision-making and implementation; make subsidies, investment incentives, or tax credits available to economic actors willing to supplant their conventional technologies with E-Ts and/or tax impositions or sanctions to correct negative externality; and educate and advertise communities to achieve more local acceptance and participation. Since the 1970s and through the decades, government E-T programs have incorporated quite a great deal of these measures and have grown into a complex economic regime of governmental institutions, regulations, and policy instruments.[2]

However, government efforts to facilitate the development of conventional industries, by deregulating or regulating and/or prioritizing and/or cooperating with or delegating to targeted industries have not necessarily facilitated the development of E-Ts. Although these government efforts are made, a great deal of E-T businesses in many political economies gradually decline or follow the boom-and-bust pattern, particularly after any such government effort is withdrawn. For example,

- Duales System Deutschland AG (DSD), the centralized German packaging waste management company licensed by the German government, shortly after its creation in 1992, was heavily cash-strapped; the overdue bills of the waste-recovery firms had to be converted into long-term debt.[3] Similarly, although increasing recycling regulations and incentives have been in place in the United States, the national aluminum can recycling rate fell gradually from 65 percent in 1992, to 55 percent in 1999, and to 48 percent in 2002;[4] the PET bottle recycling rate fell gradually from 38 percent in 1994 to 23 percent in 1999;[5] the glass bottle recycling rate fell from 32 percent in 1994 to 30.5 percent in 2000.[6]
- Most American solar companies collapsed after federal solar tax credits terminated in 1985. Even though the latest government programs are emphasizing on regulatory compliance and market instruments, according to the data from the U.S. Census Bureau data, 19 out of the 22 surveyed states saw the number of their solar-heated homes substantially decline between 1990 and 2000. Kansas, Maine, Massachusetts, Nebraska, Nevada, New Hampshire, Vermont, and Washington even had more than half of their families retreating out of solar heating. Although three saw an increase, the degrees were very slight.
- While the Danish government was poised to discontinue the statutorily guaranteed fixed price for wind energy producers in the beginning of 2000, the order for wind turbine came to a standstill. Speaking to Reuters on January 25, 2002, Economy Minister Bendt Bentsen said, "We are very concerned about the costs for society and for Denmark's competitiveness if we continue to expand the use of green energy."[7] Meanwhile, in response to the suspension of

tax credits in the United States, the world's largest wind energy market, share prices in Vestas Wind Systems and NEG Micron, the then world's largest and third-largest producers respectively, drastically dropped from the index of 250–300 in April 2002 to 40–65 in March 2003.[8]

Therefore, there must be other factors beyond the above-mentioned government efforts and E-T programs that are also playing an important role. One would spontaneously argue that polluting industries are more likely to be a major factor in a country dependent on manufacturing than in a country dependent on service industries. But, as we shall see in later chapters, although the Australian Government's decision not to expand its Mandatory Renewable Energy Target from 2 to 5 percent was widely reported as "a capitulation to coal interests" (Van Tiggelen 2004: 24), its wind farm and waste-recycling projects have experienced more significant setbacks and delays than those of much more manufacturing-based countries, namely, Denmark, Sweden, Switzerland, and Germany (a country that is also the world's largest and most competitive exporter of brown coal). Similarly, the Japanese government in the less manufacturing-based Hokkaido is incapable of requesting its monopolized utilities to absorb the additional wind cost. Although a voluntary feed-in tariff program promoting renewable energy has been introduced since 1992, the Hokkaido utility refused to use expensive wind power. And it stood its ground even though the utility made some concession later in 1999, limiting the wind projects to 150 MW out of 550 MW applications. Without support from the electric power industry, the completion of the first ever people's Wind Turbine (Hamakaze-Chan) to be installed in Tomamae, Hokkaido, was put off several times until 2001 when it was supported by the community-donated "Hokkaido Green Fund."[9]

In fact, the most outstanding characteristic of the E-T industry is its *conflicts of interest with conventional industries* (and other social groups if they are also powerful and their interests are endangered by E-Ts and/or related issues), leading to opposition from often powerful polluting industries to making E-T policies and to implementing their programs. Industries struggle to make immediate profits from them, and conflicts of interest on replacing conventional (and polluting) technologies with E-Ts can be a serious obstacle. This phenomenon is also reported by Schreurs (2003: 12) who notes that

there was an industry-led backlash in the United States in the 1980s against the regulatory successes of the environmental community.

> Further, strong industrial lobbies formed and worked over the course of the 1990s to oppose the agreement and make sure that the Senate would reject the agreement. The largest such group was the Global Climate Coalition, which spent millions of dollars fighting the Kyoto Protocol. (Schreurs 2003: 15)

Indeed, over the last few decades, strong business interest groups have lobbied against, actively defied, or sought to have laws and regulations requesting industries to adopt E-Ts repealed. At worst, interest groups from polluting industry and other social groups have even sued their governments and vice versa.

Heavy and continued conflicts of interest make *E-T development an exceptional policy sector*. Though Vitols (1997: 15) echoed the view of Ziegler (1989) that the German government-industry cooperative programs, in terms of indirect-specific projects for most of the new technologies, were "very successful means for achieving specific goals within a short period of time," the German electricity industry has strongly defied and even sued the government that required them to take up expensive renewable energies, placing their relationship under tremendous strain. Similarly, though Branscomb, Florida, Hart, Keller, and Boville (1997: 4–5) argued that "public-private partnerships have encouraged technology transfer from government funded institutions to private industry and have provided market-based incentives to strengthen private investment" and though this view was shared by the report of the National Research Council (1999) in a case study on American computer technologies showing the significance of government involvement in the industry, the widespread American government-industry hostilities and litigations over emissions controls since the 1960s and the failure to achieve the goals of the then public-private cooperative "Supercar" programs have again shown that *the development problem of E-Ts is different from that of other new technologies.* Josephine Cooper, president of Alliance of Automotive Manufacturers, at a panel presentation of the 2000 Future Car Congress, Washington, DC, stated that "(manufacturers) do not like (it) when government takes the wheel in the form of mandates . . . I never met a mandate I did like."[10] In a nutshell, E-T development is heavily characterized and obstructed by conflicts of interest between the usually powerful and strongly organized polluting industries and the often vulnerable and newly established E-T market players.

With pressures from dominant and powerful industry groups intensifying in virtually every industrialized country and autonomous region (and the many failures of government efforts in E-T programs), the idea of government involvement as an indispensable key to solving energy shortage and environmental problems was replaced in some political economies with a new tide of thought opposing state involvement in the economy and advocating market orientation, liberalization, and deregulation in the early 1980s. Since then, the mainstream view on the role of government in these countries has associated national economic and E-T development with less government involvement and has streamlined their national institutional structure for higher market efficiency and effectiveness.

However, this "state-free" approach, particularly after the introduction of neoliberalism in the 1980s, has almost invariably led to failure in E-T programs, as the examples included in this volume testify, particularly the Anglo-Saxon cases. In contrast, many E-T programs with more state involvement have been making considerable progress. They thus provoke us to once again study policies of state involvement and compare them with the unsuccessful ones of the "state-free" countries.

There are some environmental economic studies arguing for more state involvement through the use of regulations or state-industry relations. Porter and van der Linde (1995) argued against the incentive-based approach that was being promoted by virtually every standard textbook on environmental economics; in its place they proposed the "Porter hypothesis" that regulation could reduce production costs and stimulate growth and competitiveness in the area of pipe-end abatement environmental technology. Meanwhile, Wallace (1995) suggested an environmental policymaking that was independent from industry's special interests and was combined with a high-quality, honest dialogue between policymaker and industry. Yet, an empirical study on the *latest* E-T developments that *require industries to transform* themselves is lacking. In other words, Wallace's case studies were too brief, especially the composition of the public-private relationship, because he had not investigated how governments dealt with the most serious tensions between corporate and national interests, especially when they did so in a dynamic manner, that is, he did not analyze how the industry can be turned over to absorb E-Ts. Jaffe, Newell, and Stavins (2001: 66) indicated that "we have reviewed existing literature on various aspects of technology policy, but there has been relatively little empirical analysis of these policy options directed specifically at the development of environmentally beneficial technology." Since then, over the past few years, finally there have been some researchers exploring the exact impact of institutions on E-T development (e.g., Isoard and Soria 2001; Johnson and

Jacobsson 2000/2002; Bergek 2002; Madlener and Jochem 2002; Barrett 2005). Yet, although these analysts have broadly identified and tackled various obstacles in the renewable energy sector, very little has been written on the specific, dynamic issue of the reluctance of vested interests to adopt new technology.

Using such a single dimension to view state involvement seems insufficient. From this perspective, how different levels of state involvement across different national institutions have impacted on national development would be worth further investigation. One of the earliest contributions to this field was the monograph by Tisdell (1979) that observed that Japan and Germany differed from the United Kingdom in the degree of central coordination of scientific activity and in the extent to which particular ministries controlled a range of research activities at the macrolevel or intersectoral level. Further comparative studies on changes in contemporary capitalism have revealed the causal relationship between domestic political economy and national development (e.g., Hart 1992, Thelen 1996; Vogel 1996; Kitschelt et al. 1999b). They established a solid framework for the exploration of different political economies in which different conflicts of interest that impact on E-Ts are produced. In particular, Hall and Soskice (2001:8–68) regard "liberal market economies" (LMEs) and "coordinated market economies" (CMEs) as different "established orders" in modern capitalist countries. They classify most Anglo-Saxon countries, namely, the United States, the United Kingdom, Australia, Canada, New Zealand, and Ireland, as LMEs, while Japan and most West European countries, namely, Germany, Switzerland, the Netherlands, Belgium, Sweden, Norway, Denmark, Finland, and Austria, are classified as CMEs (2001:19–21). Government often responds to internal and/or external challenges by adjusting its own institutional orders to facilitate or regulate private actors who react to the challenges with efforts to make changes to their behavior so as to maintain their benefits. This phenomenon of adjustment forms the focus of this book. That is, when government involvement in E-T programs and policy instruments is insufficient, state power is required to tweak its institutions.

However, although, as mentioned earlier, many studies on the institutional impact on national development have better explained many different government policy patterns and even their outcomes, they are neither arguments for resolving conflicts of interest nor arguments for a developmental model for transformative technologies such as E-Ts. In fact, the comparative institutional perspective is useful for understanding how market actors react to challenges in the context of given institutional opportunities, but not for formulating a clear concept of state power or market force to deal with the unique problem of conflicts of interest in the E-T sector. This prompts us to pursue the study of impact of state involvement on E-T development from an *additional* perspective, one that factors in the virtue of state-industry relations and their interactions.

Indeed, as Bell and Wanna (1992: 23) argue, "explanations of business-government relations need to take into account the kinds of broader structural and institutional factors." There are quite a few examples of state-industry relations and their interactions in terms of "governance through networks, associations and communities" (Bell 2002: 9–11). However, what sort of state-industry relationship can bring about more efficient diffusion of environmental technologies? Other studies on theories of state-industry relations and their interactions are available. Wade (1990), in his study of Taiwan, attributed the successful economic development of Taiwan-like that of South Korea and Japan to the state's strategic guidance, that is, the state guided both public and private industry with a development strategy and accordingly established the "governed market" (GM) concept. Evans (1995) looked into the evolution of the information technology sector in Brazil, India, and South Korea during the 1970s and 1980s to test his "embedded autonomy" (EA) concept. Weiss (1988) illustrated her concept of "governed interdependence" (GI) by comparing five countries, namely, Japan, Taiwan, Sweden, South Korea, and Germany, and drawing some relevant evidence from the United States, the United Kingdom, Australia, and Singapore. The existing ideas of GM, EA, and GI, however, are government-industry cooperative concepts based on the condition that industry is willing to listen to, obey, cooperate, or negotiate with government. As far as the impact of state involvement on dealing with defiant actors such as the often powerful and intransigent polluters is concerned, very little literature is available. As discussed above, the three perspectives, that is, "deregulation versus regulation," "LMEs versus CMEs," and "the nature of public-private relations" could be combined into more suitable frameworks for analyzing and evaluating the E-T policies in question. The studies in this book may also contribute to the suggestion of possible solutions to the actual problem concerning serious conflicts of interest that have for decades prevented the full success of national E-T policies. That is to say, scholars in related fields have extensively explored the impact of institutions on national development and the role that the government played in it; however, although most of them concentrate upon market instruments, incentives and sanctions, laws and regulations, institutional structures, government-industry cooperation (either government-guided or industry-led or government-industry negotiated one), little effort was made to settle domestic conflicts of interest so as to clear the road to better E-T policymaking, implementation, and outcome.

This book attempts to contribute to this gap of knowledge. An alternative policy strategy called "pacifying permeation" in addition to those suggested above is generalized across the case studies, showing how the principal argument in this book works, that is, how it tweaks institutions (in the shadow of the state). The word "tweaking" refers to a dynamic concept; "obliging" regulation" and "pacifying permeation" are policy options/strategies to help tweak institutions that are composed of static rules and regulations, and of public-private relations. All these terms will be illustrated in great detail by case studies in later chapters.

For this purpose, I have established three propositions to explain the relationship between three independent variables and E-T development.

First, I suggest that E-T diffusion can proceed from neither market-oriented "deregulation," nor industry-led "self-regulation," nor "coercive regulation," but rather from dynamic "obliging regulation." That is, due to national energy security concerns and pressing environmental problems, a considerable degree of "obliging" state involvement to help E-T development becomes inevitable. The word "obliging" here refers to the government that is willing to be helpful without causing panic from regulated actors, particularly the powerful industry in this case study, so that the latter can tolerate stringent regulatory arrangements (e.g., the adoption of expensive wind energy or the achievement of higher packaging waste recycling rates) *without feeling a need to fight back*. In other words, the term "obliging" is used to describe a "nonconfrontational nature" of regulation when government is forced to change its attitude toward strongly defying industry. However, the employment of "obliging" regulations does not mean that government loses its authority or power. Rather, it is one of the tactics that the government employs in the whole transformation process to achieve desired development goals. In this context, laws and regulations have to be gradually modified according to the real situation—and not according to pure market orientation and deregulation or coercive regulation—to mould targeted actors' behavior.

Second, I suggest that, while employing dynamic "obliging" regulations, governments perform in different ways due to their different state-societal structures, that is, state and business/social structures, thus leading to different policy outcomes. So it is not just the type of regulations that matter, but also the institutional structure that affects their implementation.

Lastly, in addition to "obliging" regulation and the impact of state-societal structure, I suggest that "pacifying permeation," which is a subtle form of state-societal relationship in terms of an elastic state effort, unlike the state-led governance described by GM and the four forms of state-industry interdependent governance described by GI (their details will be discussed in chapter 2), has driven E-T development and successfully penetrated the polluter-dominated society. "Pacifying permeation" signifies a dual concept of state efforts, "permeating" the polluter-dominated industry with E-Ts by push-starting the most willing or cooperative actors while "pacifying" the sufferers.

This book, in order to make comparisons across different institutional structures, takes advantage of the conspicuous E-T policy outcomes that have recently emerged in some countries and further provides an institutional account of their E-T policy trends to explore the implications of their different policy patterns and outcomes. By comparing and analyzing the impact of distinctly different institutions on polluting industries and E-T developers, the following questions can be answered: Do national governments differ significantly in their policies to promote E-T development? And if so, in what ways? Can those policy differences be attributed to systematic institutional differences, such as those typically classified as LMEs and CMEs? Are some institutional configurations more effective than others in promoting E-T policies, and if so, why? What do these effective institutional configurations look like? Or do LMEs and CMEs each have their own strengths and weaknesses? What is the role of government in these institutions? To what degree and extent have they contributed to effective policymaking and implementation for E-T development? These research questions are of crucial importance, both practically and theoretically.

First, there is a need to accelerate the pace of E-T development in many countries. For example, when the 1996 EU Directive for the Internal Market for Electricity was installed, all EU member states were required to open a percentage of electricity markets to competition, at least 25.37 percent as of February 1999 and 30 percent as of February 2000. The development of E-Ts such as renewable and other new energies have then been generally regarded as the most effective way to satisfy further requirements of EU liberalization of the energy market. In addition, although many developed countries declared their reluctance to rectify the 1997 Kyoto Protocol, the seventh Conference to the Parties to the UN Framework Convention on Climate Change in November 2000 adopted the Marrakesh Accords that set out in detail the process of implementation of the protocol and allowed domestic actions to be taken in 2005 for achieving a 10 percent reduction in CO_2 emissions by 2010, further pressing the member countries to live up to the promises and to act quickly.

Second, there is a need to further study the virtue of institutions. Hall and Soskice (2001: v) stated that "Nation-states may derive comparative advantages from their institutional infrastructure, but few theories have been developed to explain precisely how institutions generate such advantages." Based on the fact that the extent and degree of state involvement through institutions in national development has for long been a difficult and complex policy issue, studies on institutions are still on-going, particularly on their positive impact on national development and the cogent evidence it leaves. In doing so and according to the very nature of E-T development, this book continues to examine the role of government in national development by focusing on E-T policies across different countries and times. Given these practical and intellectual needs, the findings and generalizations of the research outcome that contribute to them warrant minute secrutiny.

Concepts of Institutions and State Capacity

I employ "institutions" concepts whose analytical framework attributes international differences in economic outcomes primarily to the different domestic institutional structures in order to explain the E-T policy differences among countries compared in this study; and "state capacity" concepts, which regard effective national development as a result of state power, to explain why some of the sampled countries have a more effective E-T policy outcome than the others.

Traditionally, institutions were regarded as laws—constitutions or formal organizations of government—that is, *rules* of political systems (e.g., Wilson 1956; Robson 1960; Polsby 1975). Yet, the academic concept of institutions has evolved rapidly over the last two decades. Institutions have included not only rules but *norms* (e.g., Olsen 1984: 734; North 1990: 3; Marsh and Stoker ed. 2002). The concept of institutions was even extended to a broader dimension by March and Olsen (1989: 17), who claim that "Institutions influence actors" behavior by shaping their *values, norms, interests, identities,* and *beliefs.* Meanwhile, *organizational arrangements* have gradually been adopted by some scholars as an influential part of institutions in order to explain phenomena in their studies (e.g., Katzenstein 1978; Johnson 1982; Hall 1986). Aiming at improving classic theories of comparative advantage and to better explain behavior of firms, Hall and Soskice (2001: 9) following North (1990: 3) combine most of the above features of institutions by regarding them as a set of formal or informal rules as well as organizations' rules that actors generally follow. Furthermore, Weiss (2003: 20–21) proposes that institutions as *rules* may set constraints on individual behavior; as *norms* are constitutive of interests and identity; and as *organizational arrangements* may either enable or constrain desired outcomes, depending upon the particular policy goal. In Weiss's terms, these constitute a "thick" conception of institutions including both the "hardware" and the "software" of rules and organizational arrangements. Accordingly, this book compares political economies by dividing them into two types of market economies having dissimilar institutions. While some of the most important ingredients of institutions in LMEs are markets characterized by arm's-length relations and high levels of competition, some of the most crucial components of CMEs are powerful industry associations and unions, and their extensive collaborative networks (Hall and Soskice 2001: 9). However, these definitions and divisions of institutions have five aspects that need to be clarified.

First, institutions are not timeless and static, but ongoing and dynamic, because institutions are revised from time to time particularly when government finds a problem in its political system or feels a need to overhaul laws and regulations to deal with new challenges. For example, over the last two decades, the institutions of the British power industry have continually changed and allowed consumers access to the grid and renewable energy producers. The 1983 Electricity Act sparked off the formation of an independent power production industry and the 1989 Electricity Act allowed nondiscriminatory access to the national grid. Thus, in 1990, the British state-owned electricity industry was divided into generation, transmission, distribution, and supply, all these four parts (especially of the generation part) were gradually privatized, and power supplier choices for industries introduced.

Second, there is no such entity as a "pure" LME or CME country. The comparison is merely based on the possible emergence of a "dominant" institutional pattern. The idea of institutional coherence in a national political system should not be confused with some institutional anomalies and contradictions that often coexist alongside dominant institutional patterns (Crouch and Streeck 1997: 1–18). The Beverage Container Act, which has been imposing since 1975 a compulsory statewide return/collection mechanism in South Australia, is a good example of such an anomaly existing alongside the dominant LME pattern of the other five states and two territories in Australia.

Third, the "academic" concept of institutions should not be confused with the "literal" meaning of institutions such as a large organization that is influential in the community or a large financial fund that has considerable resources to make investments. Rather, institutions in this book refer to rules (formal or informal) and arrangements of political business or social organizations, establishments that have long-term engagement with political or economic functions. For example, the Japanese Ministry of Economy, Trade, and Industry (METI) itself is a physical institution, but its established law, custom, practice, or political structure, however, comprise the kind of academic notion of institutions that I interpret and pursue in this book.

Fourth, the scope of institutions that this book refers to are those *long-term rules and organizational arrangements* that exist as an infrastructure in which a targeted industry is developed or transformed, it excludes those *short-term* rules or arrangements that government treats as a measure that is either temporary or emergency. For example, The American federal tax incentives for solar heating in the late 1970s and early 1980s were conceived as only a temporary arrangement. They ended in 1985 because subsidization was only a stopgap solution and it was not regarded as a systematic pattern of shared expectations or an accepted norm of American policymaking. This is thus treated as a weak institutional arrangement. Instead, strong institutional arrangements comprising regulations and the state and industrial/social structures and relationships will be studied in this book.

Fifth, some substantial structural arrangements though not necessarily developed by the state can also be institutions in their own right. Alternatively, they can be partly operated by the state and partly delegated to industry (or deliberately ignored) by the state, depending on the relationship between the two. For example, in Switzerland, the government in 1989 authorized its drink industry and PET plastic industry to design and operate their own collection and recycling rules and systems for PET-bottles by setting up PET-Recycling Schweiz (PRS). However, it was not right from the word go that such designing and operating amount to an act of enforcing or delegating or enacting an "institution"—because it is only *after* such rules become a regular and routine practice that they begin to resemble an institution—and the sources of successful routinization are a separate issue.

Nevertheless, institutions are fundamentally driven by state power. Even though the state sometimes allows its industries to design and exercise their own rules and regulations, a carte blanche seldom given. A considerable segment of state leadership still works from "behind the scenes." Overall, government is the sole body responsible for choosing when, whom, and even how to manage a particular policy program. This gives rise to the notion of *state capacity*. Weiss (1998: 4) defines state capacity as "the ability of a state to adapt to external shocks and pressures by generating ever-new means of governing the process of industrial change." There are three causal relationships between state power and state capacity occurring in domestic policy sectors that need to be carefully analyzed.

First, exercising state power does not necessarily result in gaining state capacity. There are several statements supporting this premise: Both the *predatory/rent-generating* type and the *incompetent* type of state power are "more likely to stifle or obstruct industrial transformation than to induce it" (Weiss 1998: 17–20); *incredibility* of officials, *irrational expectations* or *crowding in/out* of economic actors, and *undesirable equilibria* through different policy outcomes are factors jeopardizing state performance (Glazer and Rothenberg 2001: 1–3); state power works "if they serve the common good" (Scharf 1997: 153; Montpetit 2003: 6). Thus, state power may have to be exercised "in due course," "to due extent," and in a bona fide manner to avoid these problems; otherwise, state capacity cannot be obtained.

Second, in relation to "in due course" and "to due extent" of state power per se, the acquisition of state capacity may also be contingent upon the institutional structure of a particular sector. Similar degrees and extents of state power in different sectors in a country do not simultaneously acquire state capacity and do not necessarily eventuate in similar outcomes either. Their outcomes depend on their respective institutional structures. This view is shared by Glazer and Rothenberg (2001: 8), arguing through his case studies in the United States that these are "two types of behaviors that share many similarities, but where the outcomes diverge." Also, Sharman and Moon (2003: 260), through their studies on state governments in Australia, "conclude that there is no evidence of general convergence in the overall scope of state governments."

Third, amid the globalization process or the EU integration trend and its liberalization policies, nation-states have not necessarily reduced their maneuvering space in the domestic policy arena. Instead, their individual state powers through their own domestic institutional structures, rather than through the regional/international regulations or market forces, shape the regionalization or globalization process for national goals (e.g., Fioretos 2001; Vitols 2001; Wood 2001; Casper and Kettler 2001). Thus, state capacity is produced through domestic institutional structures, rather than through external pressures, to acquire national growth.

Methodological Considerations

In this book, I compare the institutions in the most significant and mature E-T sectors across different advanced industrial countries. The comparisons are made in terms of laws and regulations, state-business structures, and public-private relationships among pollutants, E-T developers, and their governments. I do this to reveal what has happened in their institutional structures and to investigate how governments transform their own industries in the area of environmental economy. Thus wind energy diffusion cases in the United Kingdom and Germany are good cases, for these two countries have generally very different laws and regulations, that is, the British market-oriented deregulation policy for its professional, entrepreneurial, and large independent power producers (IPPs) versus the German coordination-oriented regulation policy for its inexperienced, protected, and small rural wind energy producers. But, both of these two countries, under the same EU electricity industry liberalization directives and reforms to permit entry of new energy sources, have been affected over the last two decades. So the logic behind this comparison is based on the link of their most distinctively different institutional structures to different policy outcomes. In doing so, a close examination of their national regulatory frameworks will be conducted.

Further, as discussed in the section on concepts on institutions above, some institutional anomalies and contradictions often coexist alongside dominant institutional patterns. Waste recycling industries are always governed under institutional anomalies. As an exception, they have more or less the same degree/types of laws and regulations regardless of whether the country is market oriented or coordinated oriented; they are excellent samples for the examination of the causal relationship between state-societal structures rather than between laws/regulations and E-T policy outcomes. Thus, a comparative study of the institutions in the beverage container recycling markets of three Anglo-Saxon countries (the United States, the United Kingdom, and Australia) and three European countries (Germany, Sweden, and Switzerland) is conducted, because they have similar demanding regulations and systems to ensure industry compliance with rules (e.g., beverage container deposit regulations), but at the same time they have different state-societal structures (e.g., Anglo-Saxon state-private confrontation versus European state-private cooperation).

Lastly, a comparative case study of environmentally friendly car development between the United States and Japan is carried out since both of these countries are well-known for their own government-industry cooperative programs in the auto industry, for example, the American Partnership for a New Generation of Vehicles (PNGV) and the Japanese Advanced Clean Energy Vehicles (ACE Project). These two states have similar degrees/patterns of regulations and public-private cooperation in this particular area, but they operate their cooperative programs under very different state-societal relationships, that is, American legislature's powers versus Japanese bureaucratic powers.

I will thus conduct the three comparative cases studies by examining their different mixes of laws and regulations, state-societal structures, and state-societal relationships causing the outcomes in question (see their illustrations and examples in charts 1.1, 1.2, 1.3, and 1.4).

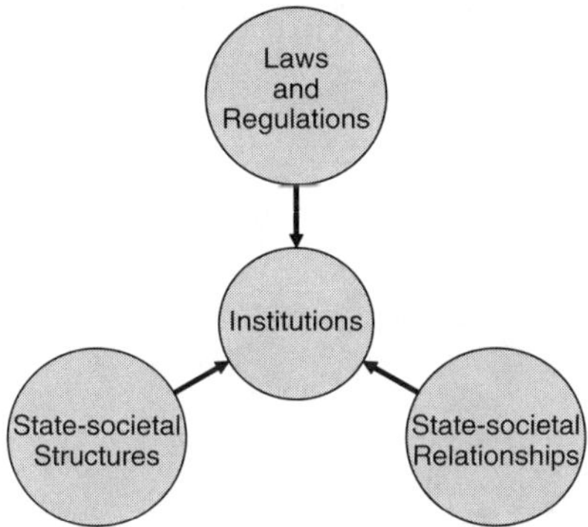

Chart 1.1   Elements of Institutions in the E-T Industry.

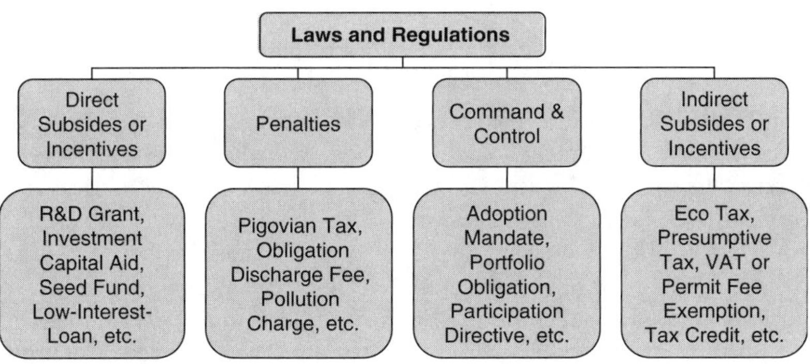

Chart 1.2   Examples of Laws and Regulations.

Note: Laws and regulations can be formal or informal, legal or nonlegal, for example, legislation and case law are legal rules; whereas Conventions and EU and Community law are informal and/or nonlegal rules. More examples will be seen in chapters 4, 6, and, particularly, 5.

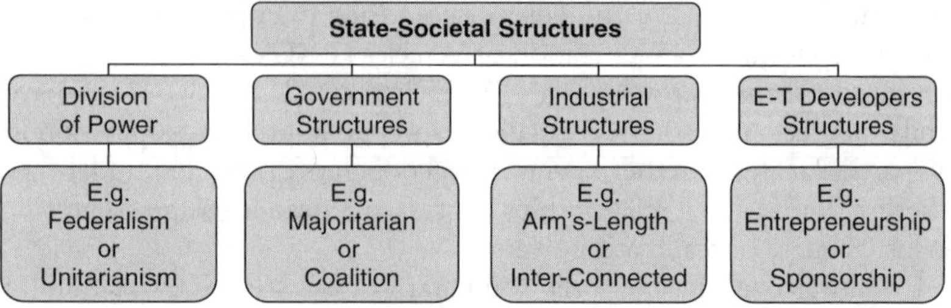

Chart 1.3   Examples of State-Societal Structures.

Note: Federalism and Unitarianism division of power, majoritarian and coalition government structures, arm's-length and interconnected industrial structures will be found and elaborated upon in chapter 5, whereas entrepreneurship and sponsorship of wind energy developers will be seen and explained in chapter 4.

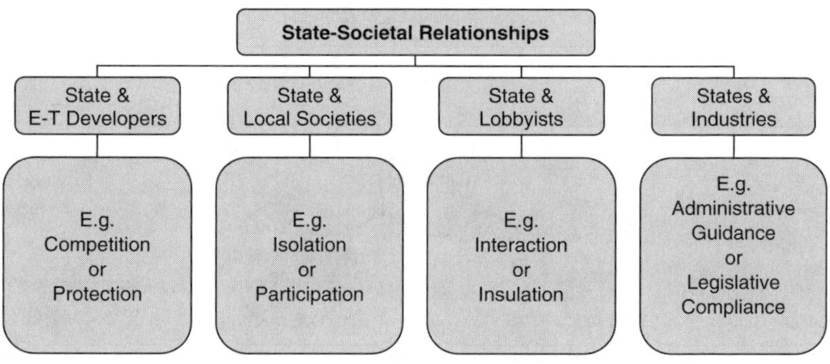

Chart 1.4   Examples of State-Societal Relationships.

Outline of the Book

This book is set out in three sections. The first section of this book has three chapters. Chapter 1 is the introduction detailing the research parameters used in the studies. Chapter 2 reviews the theoretical and empirical literature on "institutional differences" as comparative frameworks and "state capacities" as cornerstones for analyzing the role of government in E-T development. Given that governments have supported, coordinated and/or negotiated intensively with industries, the purpose of the review is to investigate whether they can also apply any or some of the existing governance concepts to the E-T development. These modes of state governance will be discussed with special attention to the serious conflicts of interest between conventional industries, E-T developers, and their governments. An important question is how state power can be successfully applied to resolve the particular development difficulties in the E-T sector, thus revealing the knowledge gap. Chapter 3 describes the research method, sources and procedures, elaborating upon a tailor-made design of "detective-like" medium-N comparative case studies. It also includes the requirements for choosing cases, and the criteria for the comparison and analyzes.

The second section of this book, from chapters 4 to 6, is concerned with the empirical findings of the case studies on wind energy diffusion, packaging recycling uptake levels, and environmentally friendly auto technology progress. These focus upon laws and regulations, and state-societal structures and relationships to explain the findings.

The last section of this book discussing results and comparing findings with previous research is presented in chapter 7, concluding with a discussion of implications and recommendations for future research.

Last but not the least, types of E-Ts and their policy categories are placed in Appendices in order that the flow of the institutional study is not interrupted. If the reader wants a brief technical description and definition of E-Ts and a general idea of the policy options available for E-T development, then it is recommended that the appendices be read before the case study chapters.

2

Tweaking Institutions in the Shadow of the State

2.1 Introduction

As discussed in the introductory chapter, domestic institutional structure is treated as the regulatory framework of a political economy molded by its state and societal structures, and by their relationship with each other. In response to government requirements, industries in those advanced industrialized countries have managed to transform their conventional technologies into environmental-technologies (E-Ts) in different ways subject to the political and social institutional structure and the relationship between the two. During the transformative process, government has been continuously asking its industries to comply with laws and regulations for adopting E-Ts. But, as posed earlier in this book, why did some industries listen to the government more willingly and fulfill similar obligations more effectively than others?

Much of the literature that has studied the E-T program per se has confined the search for reasons to either the type or the quality of government programs and policy instruments. Nevertheless, the recent emergence of more extensive use of renewable energy, outstanding packaging waste recycling rates, and the widespread sales of environmentally friendly cars in Japan and some West European countries outstripping those in the United States and some other Anglo-Saxon countries have signified that similar programs or instruments in different national political economies (i.e., different institutional structures in academic terms) may produce different outcomes in the making of a growing E-T industry. Thus, domestic institutional structure for industrial transformation becomes the centre of attention in this book. This chapter tries to answer the research questions by studying those outstanding academic articles that emphasize the significance of domestic institutions that emerged over the last 30 years and encompasses specific studies on the impact of (1) formal and informal national rules and regulations, (2) state-societal structures, and (3) the relationship between these two factors, thus providing an analytical framework for the case studies that follow.

Case studies on E-T development across different domestic institutions are concerned with these three ongoing conflicting intellectual debates. Their outcomes will form the backbone of the argument here in this book. The first contending debate is about *the role of institutions in regulation* and about the impact of institutional laws and regulations on national developments (e.g., Vogel 1996; Story and Walter 1997; King and Wood 1999; Busch 2002; Grossman 2003). While some governments have made a tremendous step forward to regulate their industries for consumers, others are afraid that regulations are unfriendly to business and that they ultimately make a mess of industrial growth. This acrimonious debate, particularly among many policymakers, is important to the study of E-T development because while some governments have promoted "deregulation" policies for many of their E-T development programs (e.g., renewable energies in the 1980s and early 1990s), others have heavily regulated and controlled their E-T sectors.

Meanwhile, new governance modes through regulatory innovation have been the subject of wide discussion in the state or reform regulatory literature (Moran, 2001a&b; 2002; Cook, Kirkpatrick, Minogue, and Parker eds., 2004; Jordana and Levi-Faur eds., 2004; 2005). They include the typical view of Levi-Faur (2005: 5) that "regulation is a necessary condition for the functioning of the market . . . regulation is helping to legitimize markets and facilitate transactions by enhancing trust." This rise of "procompetitive" or "promarket" regulation is one of the defining features of these models of a new regulatory order. This can be shared by those governments who recently promulgated substituting state discretionary power regulations for the so-called market-oriented tradable permits or certificates. However, substituting state discretionary power with better market-based/driven decisions is not the aim of such regulations. The state's regulatory role should shift from "setting down hard, *ex ante* rules and powers" to "meta-regulation," that is, making more softened, *ex post* rules and giving the state a steering role that includes "legal underpinning for indirect control over internal normative systems" wherein, nevertheless, "ends are ultimately set and determined by the sovereign state" (Scott 2004: 167–168). If light-handed regulation is to play a more important role than deregulation and heavy-handed regulation, will it really work in E-T leading sectors or countries? Do they share the view of "the making of a new regulatory order" (Jordana and Levi-Faur 2005) and/or "the rise of the post-regulatory state" (Scott 2004)?

Furthermore, for the past few decades, *direct* provision through *monopolized* and/or *fully* state-owned enterprises (e.g., electric utilities) has widely been deemed an ineffective regulatory governance mode. Procompetitive market liberalization and regulatory reform (preferably managed by an independent government agency or depoliticized bureaucracy) in many sectors-such as the polluting electricity industries in the United States and the United Kingdom, industries that have traditionally been dominated by state-owned enterprises-have replaced state-owned enterprises as part of a global phenomena (Cook, Kirkpatrick, Minogue, and Parker 2004: 3).

However, under a new regime of light-handed regulation and a steering role for the state, there could be two different general "bottom-up" directions in terms of which we could understand such a shift in E-T regulatory development. On the one hand, more *indirect* regulations have been employed. For example, more delegation of regulatory reforms to industries in some sectors/countries in which corporatism has been a major issue like the electricity industry of Germany) could have witnessed a shift in a balance of regulatory power marked by the movement away from "direct provision" toward a mode that is different but similar to "self-regulation." On the other hand, jobs of regulation-making and enforcement could have increasingly been delegated or perhaps even completely handed over to the society by the sovereign state or independent regulatory agencies "in the age of governance" (Jordana and Levi-Faur 2004/2005). For example, where state institutions have been under varying degrees of prolonged social pressure, such as those active, strong civil societies or environmental pressure groups in the United States, more of a "bottom-up" approach, than a "top-down" (i.e., the need for complying with international regulations such as the 1997 Kyoto Protocol) or "horizontal" one (i.e., the commitment toward or desire for catching up or joining forces with peer countries, such as the implementation of environmental work plans in the UN or OECD) has been evident (terms quoted from Levi-Faur 2005: 8) in industrial transformation.

Yet, "bottom-up" approaches in the advanced industrial countries do not necessarily refer only to Levi-Faur's (2005: 8) description of "a new division of labor between state and society" or "self-regulation in the shadow of the state." It may be entirely due to other factors since primary evidence is lacking on how the "bottom-up" approach can be realized, functionally and effectively, for achieving E-T policy goals. The "bottom-up" approach in countries dominated by strong, polluting industries or by the proenvironmental civil societies could refer to different specific analytical terms. For instance, while the approach in the United States would be related to the mode of "bottom-up regulation" controlled by polluting industries or by environmental pressure groups to prevent legislators from making "unfavorable" laws and regulations, the approach of the German government would be to pet in place undergone a "bottom-up" process of regulatory reform through a coalition "effort to promote bottom-up exchanges and to reach consensus on both problems and solutions" (Doner and Ramsay 2003: 136). After all, do the "top-down," "horizontal," and "bottom-up" modes of regulation or processes of E-T regulatory reform—or both—represent a better alternative of governance than others under the circumstances? In this book the shift between the approaches ("top-down," "horizontal," and "bottom-up") is explored by comparing empirical data form E-T sectors of selected liberal market economies (LMEs) and coordinated market economies (CMEs) whose processes of E-T policy reforms are heavily shaped by industrial/social involvement and/or regulatory mode in different contexts.

The second contending debate is about the "breadth" and "length" of regulations. That is, other than political regulations, do other types of regulations such as *social norms* or *state structures* also have an impact on national development? Besides the "institutional laws and regulations" approach as mentioned above, there is an interesting array of literature on the impact of institutional norms and/or structures (e.g., Johnson 1982/1984; Hall 1986; Nester 1990; Hart 1992; Evans 1995/1998; Hall and Soskice 2001). In this debate, while some scholars are doubtful of the broad definition of institutions (e.g., Rothstein 1996: 145), others suggest that institutional norms and/or structures can better regulate patterns of behavior and they regard them either as complementary or even as a factor more important than institutional laws and regulations (e.g., Goodin 1996: 22).

Furthermore, the serious market and systemic failure of E-T development has often necessitated the intervention of state power and government involvement. On the role of the state and the public-private relationship, scholars' views range from those that emphasize the role of authoritative power over industries, to those that identify the importance of negotiated power in cooperation with industries. Since national development is not timeless but dynamic, no scholar would believe his/her theories to be applicable to all periods. A study of how to exercise state power through institutions "in due course" and "to due extent" is thus also very important. So the third ongoing debate concerning the public-private relationship in the national economy is also taken into consideration. State governance concepts or theories in terms of government-industry relationship and policy strategy have been steadily formulated over the last two decades (e.g., Mann 1984/1988; Wade 1990; Hart 1992; Evans 1995/1998; Weiss 1995/1998). It would be important to examine public-private relationships at different stages using these different concepts and theories.

Over the last 30 years, studies on domestic institutions concentrated their efforts on different economic affairs. As we shall find later in this chapter, each of these studies represented a common effort to explain contemporary economic performance. The first line of debate for these studies began at the outset of the era of economic neoliberalism in the early 1980s. It offered abundant evidence of the different arrangements between "deregulated" and "regulated" institutions, leading to remarkably different outcomes. Meanwhile in Western Europe, EU integration has provided an abundance of interesting comparative data between the neoliberal economic structure of some European countries such as the United Kingdom, and the coordinated economic structure of other European countries such as Germany. These distinctly different data have attracted many economists to engage in institutional studies.

Meanwhile, on the other side of the world, Japan's "economic miracle" in the postwar period and the rise of those Asian "Tigers" and "Dragons" have aroused the interest of many scholars to look at specific Asian-style institutional structures. Naturally, the noticeably unique values and traditions in East Asia are propitiously taken by many scholars as new institutional sources. Comparisons of national institutions made in a broader sense rather than those made only between "deregulation" and "regulation" have steadily emerged.

Yet, while this group of broad institutional comparisons has concentrated on the causal relationship between the characteristics of static state-societal institutions and national performances, the liberal market policymakers and economists are concerned with the question of whether regulations, coordination, and strong state powers for E-T development are harmful and more costly to industry. In turn, these would be detrimental to the national economy and thus would provide a strong reason to turn down government intervention in the national transformative process. By contrast, coordinated market policymakers and economists have given primacy to more laws and regulations and/or emphasized the influence of the normative features and/or the organizational arrangements of institutions. They recognize that benefits lie not in the immediate maximization of national wealth but in the strategic fulfillment of national goals. As this concept achieved greater significance in terms of national development in many countries, particularly during the postwar period, academic and policymakers engaged in an untiring search for their clarification.

Another group of scholars has been interested in the formulation of more dynamic state power concepts or theories such as "infrastructural power," "governed market power," "embedded autonomy," or "governed interdependence." The studies included state structures, state-industry relations, and, most importantly, how they shape possibilities for industrial transformation amenable to E-T development.

I begin with a general debate on the role of the state in regulation and follow it up with an analysis of the studies on national structural differences and relevant concepts and theories of state-societal relationships. Additional scrutiny will determine whether these studies, concepts, and theories can furnish a satisfactory explanation why E-T diffusion is effective in some political economies and not in others.

2.2   The Role of Institutions in Regulation

After experiencing the pros and cons of regulatory policy during the Great Depression and World War II and Keynesianism (which was increasingly replaced by supply side policies) in the postwar period, the long drawn out debate between market forces and state powers has evolved into another debate that is highly influenced by the literature of neoliberal economics (based on the concept of the rejection of government intervention of neoclassical economics that emerged in the late nineteenth century) by, for example, Friedrich von Hayek and Milton Friedman in the mid-twentieth century.[1] Meanwhile, the "public interest" school of regulation in the 1950s and 1960s interpreted regulation solely as a means to correct market failure problems such as imperfect market information, natural monopoly, or negative externalities. In addition, while some economists assert that the larger the market pursued, by business actors is, the higher the potential for improved productivity, competitiveness, and profits (e.g., Scitovsky 1960: 282–283), others believe that the smaller the market managed by government is, the easier it is for a society to regulate and exploit the development of modern technology (e.g., Kuznets 1960: 28–32). In any case, the debate over regulatory issues entered into a new era when economic neoliberalism was finally adopted by the Anglo-Saxon regimes and drove the process of globalization in the 1980s.

Scholars have since studied and classified government policies into "deregulation" and "regulation." The general term "deregulation" refers to a state policy that the state removes regulations, control, and/or other barriers that may restrict the development of an industry. "Deregulation" tends to forge a competitive economic climate by dismantling regulatory barriers to businesses. However, under the principle of neoliberal economics, "deregulation" does not necessarily denote a complete *laissez-faire* but rather the construction of a competition arena through the introduction of changes such as the termination of monopolies in the electric power industry. Studies on "deregulation" have mushroomed in many advanced industrial countries since the "deregulation" policy was widely implemented over the last two decades as a main economic reform. By looking at the first decade of neoliberal reforms, quite a few economists have held the opinion that regulation reduces economic efficiency (e.g., Peltzman 1989; Noll 1989; Duch 1991; Niskanen 1993). In addition, some scholars have also suggested that advanced information technology and cross-border electronic transaction, particularly in the financial and banking sector, have greatly weakened the impact of regulation (e.g., Wriston 1992; Ely 1993); and severe global competition has encouraged governments to remove most of their regulations (e.g., Cerny 1991; McKenzie and Lee 1991). Overall, scholars at the time suggested that the powers of government were being reduced by market forces due to the increasing impact of economic neoliberalism and globalization (e.g., Swann 1988; Naisbitt and Aburdene 1990; Toffler 1990; Ohmae 1990).

Although the findings of these scholars supported less regulation and more market competition, "deregulation" in the E-T industry has, however, led to an unsatisfactory outcome. For example, early wind power policy in the United States did not succeed because the federal government was prevaricating between protectionism and neoliberalism without considering regulations to remedy the actual development problems. During the Carter administration of the late 1970s, wind power projects were initially encouraged through the Installation Tax Credit (ITC), a temporary incentive offered by the federal government's 1978 Public Utilities Regulatory Policies Act (PURPA). From 1978 to 1985, the ITC provided a tax credit for electricity generated by wind turbines. The value of the ITC was based on the installed cost of a project and, however, not on performance. In addition, the application for the ITC was free of regulation. As a result, the ITC caused a wind turbine rush (e.g., approximately

1,600 MW of wind generation capacity or some 1,000 turbines were ordered in Palm Spring Valley and the valley on Interstate 10 in the early 1980s), but its growth was not really demand driven.[2] Hence, the next federal government managed by Reagan's neoliberal reformists withdrew the ITC in 1985. Since then wind farm installation has retarded significantly and most American wind power producers and turbine manufacturers shut down in the next few years.

Meanwhile, although very few scholars disagreed on institutional rules and regulations (e.g., Borrus et al. 1985; Hills 1986; Moran 1991/1992), the debate in the early days was conducted in a far more unbalanced manner. Yet, with more and more data on market in the 1990s, more organized empirical studies on the positive impact of institutional rules and regulations have appeared, these studies show that "regulation" has not really been abandoned by governments in advanced industrialized countries (e.g., Vogel 1996; Busch 2002; Grossman 2003). The term "regulation" refers to a state policy that the government deliver rules and regulations to control or direct the development of an industry. These rules and regulations are those issued by the federal or state government and have legal force through the process of institutionalization that specify standards or procedures for implementing the government's policy programs. To develop national economy by means of rules and regulations, government officials through their institutions control the desired behavior and review conformity to them. Erection standards, project monopolies, or licenses are examples of such regulatory regimes.

Of these new arguments for the important role of institutions in regulation, Steven Vogel (1996) has viewed regulatory reforms from an institutional point of view. Based on the state institutional approach in literature of scholars such, as Shonfield (1965), Katsenstein (1978), Zysman (1983), and Evans, Reuschemeyer, and Skocpol (1985), Vogel examines different countries through case studies on technologies for telecommunications, financial services, broadcasting, transport, and utilities, in order to investigate how governments "deregulate" or "regulate" their industries. Despite a similar need to transform their industries in the era of globalization, these case studies reveal that governments (e.g., the British government and the Japanese government, not their industries) have not converged in a common trend toward "deregulation" but have instead adopted liberalization and "re-regulation" in markedly different degrees and developed "distinctive new styles of regulation" (Vogel 1996: 4, 265–268).

This view has been endorsed more recently by further studies into other policy sectors. One of these is a paper authored by Andreas

Busch (2002) who analyzed the role of institutions in European banking regulation. Such an empirical study on the role of the state in regulation through political institutions, a role tested in the banking sector, was also conducted by Emiliano Grossman (2003) on the EU level. Notwithstanding the need for regulation or so called re-regulation, a new light-handed and/or ex post facto mode of regulation is of late being, discussed by a group of scholars (Moran 2001a&b; 2002; Cook, Kirkpatrick, Minogue, and Parker eds. 2004; Jordana and Levi-Faur eds. 2004; 2005). The rise of "the regulatory state" may have been a necessity for choosing a direction further down at the crossroads of more and less heavy regulation.

The latest findings have vetoed the abrogation of regulations, and free competition by industry without regulation by political institutions, two factors that had been seen as the best ways to increase economic welfare during the era of popular economic neoliberalist reform in many countries. E-Ts are a new industry to test these latest findings. The case studies of wind energy in the "deregulated" United Kingdom and "(self-) regulated" Germany in a later chapter may provide some implications.

2.3 The Impact of State-Societal Structures

Another line of research on institutions is into the impact of social elements (e.g., social norms or organizational arrangements rather than laws and regulations) on national development. Pioneered by the early studies on societal organizations beyond state structure (e.g., Katzenstein 1978; Krasner 1978), societal institutions have begun to draw the attention of political scientists. Obviously, this line of research was touched off by the unique state-societal culture in Japan when the rapid economic rise of this mysterious Asian country after World War II surprised many orthodox economists. Thereafter, a considerable volume of studies on Japan's special economic model has been conducted, labeling the Japanese business networking and exported-oriented model as either corporatism or neomercantilism (e.g., Pempel 1977; Schmitter and Lehmbruch 1979; McCormick and Sugimoto 1986; Woronoff 1988; Wolferen 1989). Nonetheless, one of the earliest empirical attempts at scrutinizing state-societal institutions is Chalmers Johnson's *MITI and the Japanese Miracle* in 1982, which attributed the rapid economic growth of Japan to the efforts of the Ministry of International Trade and Industry (MITI or METI today). He pointed out that the nature of the Japanese institutional structure was very different from that of the United States, leading to the two countries following very different economic policy patterns. Most significantly, he highlighted the special attributes of the Japanese domestic institutions in his study. The core attributes of Japanese institutions can be seen in his introductory chapter wherein he says that

> although it is influenced by pressure groups and political claimants, the elite bureaucracy of Japan makes most major decisions, drafts virtually all legislation, controls the national budget, and is the source of all major policy innovations in the system.

In addition to the MITI's regulatory system, Johnson also appreciated the influential impact of other Japanese institutional features. These were its lifetime employment system, seniority wage system, personal saving system, distribution system, the *amakudari* (the bureaucratic elites), the *Keidanren* (The Federation of Economic Organizations) and enterprise unionism, the *keiretsu* (massive business groups) and their sub-contracting system, the debt-oriented financial system, and so on. Yet, his argument was highly based on the significance of bureaucratic power in the prewar and postwar periods (1925–1975) during which both the government and industry had commonly and strongly agreed goals through conversations between their bureaucratic institutions (such as the MITI) and societal institutions (such as the *Keidanren*) to rebuild the country from devastation, concentrating all efforts on catching up with the developed countries. This created an impression to his readers that a country's economic success can be materialized by state institutions that strongly guide the economy through societal institutions.

Johnson's views on such a bureaucracy-oriented, state-societal structure were later adjusted by Nester (1990). In his book *The Foundation of Japanese Power*, Nester (1990: 15) does not endorse Johnson's institutional views that the rapid Japanese economic growth during the postwar period was mainly caused by Japan's elite bureaucrats, even though the MITI made nearly all the major decisions in the country during that period, and policies under this sort of state were plan rational, rather than market rational, and beneficial to privately owned enterprises (especially those that were capable of incorporating new technologies and were committed to national development goals). Having observed the latest institutional development in the 1980s during which both the Japanese economy and its industrial sectors had become very strong, Nester argued that the Japanese bureaucracy was increasingly losing its monopolized powers while other actors such as large corporations were increasingly having a large say and bargaining power with the bureaucrats. He justified his arguments by elaborating the increasing domestic and foreign political power of Japanese business associations, and horizontal and vertical *keiretsu*. Nevertheless, he did not mean to completely negate the importance of bureaucracy but sparked off an expanded debate on various issues, from Johnson's emphasis on bureaucracy-led, state societal structure to a more balanced state-societal structure subject to the institutional interaction between state and societal actors.

Further to this concept of state-societal structures, Jeffrey Hart (1992: 31–33) has also shared Nester's state-societal institutional views, rebutting Johnson's and others' statist opinions (e.g., Ikenberry, Lake and Mastanduno 1988; Wade 1990). Further to Hall (1986: 17), Hart (1992: 32) criticizes that the autonomy of the "strong" state in reallocating resources across industries is not the whole story, but bureaucrats, even those in Japan, have to confer with their societal actors. His argument is based on his studies on the state-societal institutional structure, differentiating and comparing varieties of state-societal arrangements among state, business, and labor. The case studies in his book *Rival Capitalists: International Competitiveness in the Unites States, Japan, and Western Europe* (1992) have explored useful insights by looking at steel, automobiles, and semiconductor industries in a wide range of countries. Hart (1992: 1) argues that the relationships between state, business, and labor, especially during technological transitions, can accelerate or impede the development and diffusion of technological innovations and is crucial to competitiveness. Hart (1992: 281) reveals that relatively speaking France has a "strong" government but poorly institutionalized labor and industry, while Germany has a "weak" government but strongly institutionalized labor and industry; further, Japan has a strongly institutionalized bureaucracy and industry despite "weak" labor, while the United Kingdom has poorly institutionalized bureaucracy and business though "strong" labor and industry; finally, the United States has "strong" business but poorly institutionalized government and labor. Hart (1992: 283) concludes that government-business-labor relationships have constructed different domestic institutional structures among countries, leading to the outcome that both the United States and the United Kingdom had "the least successful pattern when compared with all the other large industrial countries." The "weak" state-societal institutional linkage of these two less-performing countries implies that effective state-societal structures are vital for national development.

Another line of state-societal institutional study is based on a *broad* view of institutions. In addition to the formal institutional rules, Hall

(1986) devised a concept of "standard operating procedures" that refer to informal institutional rules such as those unwritten political practices or settings that generally guide actors' behavior. Nevertheless, North (1990: 83) further expands the interpretation of informal institutional rules to tradition, custom, culture, and habit. Following this, Hall and Soskice (2001: 9), in their edition of *Varieties of Capitalism: The Institutional Foundations of Comparative Advantage*, define institutions as

> a set of rules, formal or informal, that actors generally follow, whether for normative, cognitive, or material reasons, and organizations as durable entities with formally recognized members, whose rules also contribute to the institutions of the political economy.

Hall and Soskice (2001: 1–68) in their introduction elaborated the remarkable institutional differences between liberal market economies (LMEs) and coordinated market economies (CMEs) and analyzed how corporate behavior was influenced by different institutional infrastructures, such as education and training systems, management-labor relations, intercompany relations, and corporate governance systems, of political economies. The term LMEs is elaborated by Hall and Soskice (2001:8):

> In response to price signals generated by such markets, the actors adjust their willingness to supply and demand goods or services, often on the basis of the marginal calculations stressed by neoclassical economies. In many respects, market institutions provide a highly effective means of coordinating the endeavors of economic actors.

By contrast, the term CMEs too is described by Hall and Soskice (2001: 8):

> In CMEs, firms depend more heavily on non-market relationships to coordinate their endeavours with other actors and to construct their core competencies"; they "entail more extensive relational or incomplete contracting, network monitoring based on the exchange of private information inside networks, and more reliance on collaborative, as opposed to competitive, relationships to build the competencies of the firm.

Hall and Soskice (2001: vi) further state that "building on the distinction between CMEs and LMEs is central to the approach; the contributors in the book show how firms develop corporate strategies to take advantage of the institutional support available in any economy for particular modes of coordination, deriving from this a new perspective on issues in strategic management"; and point out that "coordination-oriented policies will be more difficult to implement in LMEs because their business and labour associations usually lack the encompassing character (i.e., highly organised societal institutions) required to administer such policies well . . . If regimes that provide structural influence to encompassing producer groups find it more feasible to implement coordination-oriented policies, while states in which power is highly concentrated have more success with market-incentive policies" (Hall and Soskice 2001: 48–49).

State-societal structures are thus regarded, in a broad sense, as institutional infrastructures typically represented by the institutional characteristics of LMEs and CMEs in which firms react to them. Sixteen scholars in this book published their findings in thirteen articles, taking this approach to comparing different institutions and market economies on different themes, having tested Hall and Soskice's (2001: 12) postulation that "deliberative institutions can enhance the capacity of actors in the political economy for strategic action when faced with new or unfamiliar challenges." Still, Hall, and Soskice (2001: vii) in the book present their interpretation not as a conclusion of the findings but as an attempt to create research agenda and call for new papers to compare institutions in multiple sectors. The E-T sector may therefore provide a new area for the purpose.

This "LMEs versus CMEs" approach (which paid attention to the differences of four corporate subsystems, that is, education and training, finance, labor, and intercompany relations) to institutional analysis is a suitable prospective for individual firms in many industries. However, it is not a perfectly relevant analytical framework for most E-T development cases in which much of the story is about the power struggle between government and industry (and other social groups). The economic and sociopolitical problems encountered by the government in the E-T sector are such that they have to remove various procedural barriers and motivate industries. Therefore, state-societal institutions made up of mainly government and industries (and other social groups), and state-societal relationships that highly influence policymaking and corporate behavior, will also be scrutinized in the following section.

2.4   The Effect of State-Societal Relationships

Indeed, the above general concept of state-societal institutional structure has neglected the quality problem of the state-societal relationship. Jordan and Asford (1993: 32) indicate that when a state-society consensus is obtained, bureaucrats are enabled by the supreme power to make decisions on policies that give rise to highly efficient and effective policies, but sometimes this kind of power may cause corrupt bureaucracy and ineffective autocracy. Calling attention to this issue, Peter Evans (1995) has studied different kinds of state involvement in industry for the improvement of state autonomy. In the book *Embedded Autonomy: States and Industrial Transformation* (1995), Evans examined the evolution of the information technology sector in Brazil, India, and South Korea during the 1970s and 1980s to investigate how state involvement can equitably affect a country's place in the international division of labor. According to the findings, Evans (1995: 12–13) suggests that Korea can meet the requirements of the paradoxical concept of "embedded autonomy" that refers to a bias-free bureaucratic decision-making ability despite their regular, intimate contacts with the industry, but "Brazil and India are definitely intermediate cases." Evans (1995: 11) claims that

> [states] vary dramatically in their internal structures and relations to society. Different structures of state-industry relationships create different capacities for action. Structures define the range of roles that the state is capable of playing. Outcomes depend both on whether the roles fit the context and on how well they are executed.

He differentiates varieties of state structures and state-societal relations between the "predatory state" and the "developmental state" and attributes the success of state involvement to the virtues of a concrete set of social ties that binds the state to society and provides institutionalized channels for the continual negotiation and renegotiation of goals and policies, and for a sense of corporate coherence. Notwithstanding the "close" state-societal institutional relationship, he emphasizes that the separate roles of regulators and producers are vital for the outcomes of national transformation and he conceptualizes such paradoxical public-private relationships as "embedded autonomy," that is, bureaucracy has an intensive contact with society but can still exercise a sober judgment for making decisions and implementing policies. If an "embedded autonomy" exists, Evans (1995: 250) suggests that "uniformly treating bureaucrats as lion fodder is a mistake. Disdain is often deserved, but state bureaucracies can also be homes to creative of entrepreneurial initiatives." Yet, based on the features of the East Asian state's internal organization, Weiss (1998: 49) extended Evans's idea that "embedded autonomy" is produced not only by bureaucratic coherence through meritocracy and personal networks but also by "the state's key policymaking agencies ... sufficiently insulated against special interest groups and clientelistic pressures generally." Thus, "embedded autonomy" is also a key concept for E-T policy development where vigorous lobbying by special interest groups is predominant.

Further to the concept of "embedded autonomy," another inspiring attempt at state-societal institutional relationship by Evans was his edition of *State-Society Synergy: Government and Social Capital in Development* in 1997. He edited the empirical contributions of five scholars (namely, irrigation associations in Taiwan, industrial workers in Kerala of India, coproduction of urban infrastructures in Brazil and coproduction of primary education in Nigeria, political construction of social capital in rural Mexico, and economic development in Russia and China), arguing that public-private "synergy usually combines 'complementarity' with 'embeddedness' and is most easily fostered in societies characterized by egalitarian social structures and robust, coherent state bureaucracies." Evans (1997: 179) described "complementarity" as "the conventional way of conceptualizing mutually supportive relations between public and private actors. It thus suggests a clear division of labor, based on the contrasting properties of public and private institutions." However, difficulties emerge when we use the concepts of the above literature to resolve the development problems in the E-T industry. For example, given a wide variety of possible environmentally friendly autotechnologies requiring *different* and *conflicting* public infrastructure (e.g., liquefied petroleum gas (LPG), compressed natural gas (CNG), or electric battery refilling/recharge networks), how could governments *complement* the enormous and incompatible public goods of egalitarian social structures as discussed by Evans? In addition, "complementarity" with "embeddedness" works well by rounding off the private actor's deficiency with governmental efforts only when they have the same goals. Norms of trust and social capital from intimate partnership interaction are built up only when they have common interests. What if they have different goals and/or distinct interests: a possible scenario in instances wherein there is disagreement between environmental improvement and corporate profitability? This will be answered in the case studies discussed later in this book (chapter 6) particularly by testing whether the notion of "complementarity" would work. In any case, using the concepts as discussed above, state-societal institutions "embedded" by "autonomous" bureaucrats may "complement" societal actors. However, to what degree and extent can a state motivate its societal actors through institutions? This takes us firstly to investigate whether the state can strongly exercise its authority or compromisingly cooperate with industries. This makes us recall the distinction made between the concepts of "despotic power" and "infrastructural power" by Michael Mann (1984/1988: 1–32). While "despotic power" refers to a coercing force exercised by a "weak" state such as an imperial regime to extract the resources it requires from its society, "infrastructural power" refers to state power generated by a "strong" state such as bureaucratic government to permeate and extract resources from society through an institutionalized, cooperative, and interactive state-society relationship. "Infrastructural power" is doubtless a useful concept to understanding how a state can improve its *penetrative-extractive* capacity without resorting to the *arbitrary* use of power; however, it is still difficult to explain thoroughly how a state can achieve successful national performance (Weiss and Hobson 1995: 4). Actually, in what way can a government utilize its state power to achieve desired national performance?

One of the earliest findings on the causal relationship between state power and national performance was the *guiding* type of state power endorsed by Robert Wade's (1990) lessons from East Asia in his book *Governing the Market: Economic Theory and the Role of Government in East Asian Industrialization*. In this Wade put forward a "governed market" (GM) theory denoting that a "strong" state's governing role and its strategic industrial policies are central to national development. Wade further suggested that superior economic performance can be achieved through a strong state approach by showing "causal" relationships between the strategic industrial policies and the exceptional economic achievements of Japan and the newly developed countries (NICs). Wade (1990: 26) stated that

> the superiority of East Asian economic performance is due in large measure to a combination of: (1) very high levels of productive investment, making for fast transfer of newer techniques into actual production; (2) more investment in certain key industries than would have occurred in the absence of government intervention; and (3) exposure of many industries to international competition, in foreign markets if not at home . . . Using incentives, controls, and mechanisms to spread risk, these policies enabled the government to guide-or govern-market processes of resource allocation so as to produce different production and investment outcomes than would have occurred with either free market or simulated free market policies . . . the policies have been permitted or supported by a certain kind of organisation of the state and the private sector.

The findings of Wade obviously supported "strong" state power not only through privilege or industrial policies but also through an institutional "top-down" approach. Aiming at promoting a particular industry, "strong" states such as Taiwan during its postwar period set up helpful institutions serving target industries that were in line with industrial policy. Such institutions perform their coordination functions mainly in three ways: financial backup or strategic investment, corporate restructuring or upgrading, and R&D assistance, by knowledgeable bureaucratic elites guiding their societal actors.

Nevertheless, the GM concept could not explain conditions in a highly mature society, that is, after those developing industries have grown up and become powerful. At this moment, societal actors are not prone to follow state decisions anymore and bureaucrats also have difficulty avoiding governmental failure without gearing themselves to the society, particularly at the crossroads of national transformation. Pinpointing the weakness of GM, the innovative ideas of Weiss and Hobson (1995: 7)-who refined the concept of institutionalized state power by breaking it down into "penetrative," "extractive," and "negotiated" powers-especially the idea of collaboration between government and society rather than that of coercion of society into complying with state regulations. Highlighting the importance of "negotiated" power, Weiss (1988) in her subsequent work, *The Myth of the Powerless State*, not only suggests that the impact of economic challenges on national development depends largely on static institutional structures but she also makes the very nature of dynamic state power clearer by explaining it in greater detail than the others. This expands further the dimension of state-societal institutions. Weiss (1998: 48) reviewed Wade's GM theory again by holding the view that effective policies are performed by "regular and extensive consultation, negotiation and coordination" between a strong state and a strong industry, rather than by an industry reacting obediently under strong state pressure as described by Wade. Weiss (1998: 5) regards such negotiated public-private relationships as an effective source of a state's transformative capacity, that is, "the ability to coordinate industrial change to meet the changing context of international competition," which can "augment a society's investible surplus." She points out the weaknesses of the coercive sense of "statism" and highlights the strengths of the dynamic notion of state-society interaction that refers to a negotiated relationship between a "strong" state and highly organized economic groups. The value of such a dynamic and interactive state power in place of a static and unilateral one must be appreciated. She suggests that this sort of state power can provide states with a strong transformative capacity and help states effectively move their domestic industry toward global competition and that there are three prerequisites for achieving a strong transformative capacity: first, the bureaucracy has to be efficient and devoted to organizational objectives; second, a strong intelligence-collection infrastructure has to be installed; third, insulated pilot agencies for the coordination of industrial change have to be in place (Weiss 1998: 54). Weiss tested her propositions by comparing and macroanalyzing five countries, namely, Japan, Taiwan, Sweden, South Korea, and Germany, and drawing some relevant evidence from the United States, the United Kingdom, Australia, and Singapore.

With the support of this evidence, Weiss has theorized her findings. While Wade ascribes the relatively effective progress of early industrialization to the concept of the "hard" or "authoritarian" state, Weiss finds evidence that the more developed and complex the economy is, and the more organized the industry sector, the more the need for public-private coordination is, and the less the need for top-down control. This public-private coordination is centrally governed and interdependent, what she calls Governed Interdependence (GI). The GI concept denotes that an extensive, regular, and interactive negotiation system between the government and industry plays an important role in generating national transformative capacity. Weiss (1998: 38) defines this concept as follows:

> GI refers to a negotiated relationship, in which public and private participants maintain their autonomy, yet which is nevertheless governed by boarder goals set and monitored by the state. In this relationship, leadership is either exercised directly by the state or delegated to the private sector where a robust organizational infrastructure has been nurtured by state policies. GI is intended to convey a reality in which both state and dominant economic groups are "strong": i.e. the state is well insulated and industry is highly organized and linked into the policy-making framework via a robust negotiating relationship.

However, the existing ideas of GI are government-industry cooperative concepts that work on the condition that industry is willing to cooperate or negotiate with government. As far as the impact of state involvement on dealing with defiant actors, such as often powerful and intransigent polluters, is concerned, less literature is available.

2.5 INSTITUTIONS FOR ENVIRONMENTAL-TECHNOLOGY DEVELOPMENT

Other than for the industrial developmental purposes in Japan, Germany, and the NICs, GI may be a useful concept also for E-T development if there is a possibility to remove a priori the conflicts of interest. The negotiated form of GI is not feasible for E-T businesses from the very outset of national transformation. Even though government officials have made strong efforts to enter into agreements with industries, there is often opposition or unwillingness to negotiate with governments who ask for compliance or cooperation in E-T projects. Still, as quoted in the introduction of this book, Schreurs (2003: 12–15) has found compelling evidence of an industry-led backlash in the United States over the past two decades against the regulatory successes of the environmental community.

In the following chapters, this industry-led backlash will be illustrated also by the heavy industrial resistance to comply with regulation and the adamant reluctance to negotiate with government in the early days.

So why has this phenomenon emerged and continued? The first explanation rests on the fact that the powerful, polluter-dominated industries are utility maximizers against the less-profitable or even loss-making E-T businesses. Strongly organized civil societies such as business associations or trade unions, established to protect their own industries, and highly internationalized industries, for example, the auto-making industry, are able to powerfully defy the government, even though the latter has ardently promoted the advantages of E-T economy. E-Ts are normally replacements for conventional industries. However, as costly E-Ts jeopardize profits, especially in the face of market competition, industries are usually very reluctant to absorb the additional cost. Besides, since most industries cause pollution and are usually the basis for the "sustainable" development of a national economy, it increases the bargaining power of industries to reject government requests for adopting E-Ts, especially in those highly democratic countries where policymaking is accessible to nongovernment legislators or even common citizens. What is more, in addition to the physical reasons, it is difficult to change quickly their well-established business routines or familiar patterns of profit-driven corporate behavior. Thus, these "calculus" and "cultural" reasons are concurrently countering reform.[3] The industrial actors, especially those in the free-market-oriented countries, do not believe that the industrial policies that commonly prioritize particular industries can cover the whole industry in the context of pollution and energy shortages. This is because most governments lack sufficient political and economic resources to deal with and provide the dominant industry with greater degrees of confidence or assurance about the future profitability of the E-T economy. Thus, industries are unwilling to develop strategic long-term plans at the risk of short-term profits. Rather, they tend to maximize the attainment of their short-term goals and preferences. Moreover, many E-Ts are too localized and scattered, for example, the personal use of solar panels happens only in areas where sunlight is strong, or wind farms generate power only in remote windy places. These E-T adopters are usually inexperienced, dispersed, and small. It is not practical to form strong associations to negotiate with government, let alone get the government to delegate control over the implemention of policy programs. Thus the negotiated form of GI is not feasible for E-T businesses from the outset of national transformation.

One may suggest that if fragmented E-T industries do not allow for quality public-private negotiation and cooperation, then resource reallocation or a less commanding version (mostly by resources reallocation and financial induction) of GM might suit them better. But policymaking for E-T development is far more complicated and difficult than that of the export-oriented or developmental states. Being a catching-up or developing country over the last few decades, Japan and the NICs could easily have followed the successful path of developed countries, studied trends in technological change in various industries, import and export composition, and industrial structure (Wade 1990: 334–335); or taken advantage of the U.S. trade and political needs (e.g., the U.S. containment foreign policy supporting friendly regimes economically). However, E-T is a very new sector without any previous data and nobody can set a good example of it or support other countries to do so. If E-T development is purely reliant on the initiative of bureaucrats, there is no guarantee that successful technology initiatives can be made (examples such as the abandonment of decade-long plutonium development or pure electric vehicles are abundant), even though career technocrats or outsourced specialists have tried to ensure a neutral, sober attitude toward the search for the best technologies. This primary problem indicates that a dictatorial GM is not a good solution for E-T development either.

The concept employed here to view the rapid growth of those E-T economies differs from the "top-down" form of the GM theory or the negotiated public-private cooperation of the GI concept. According to evidence obtained from case studies, although some of the facts agree with Weiss when she argues the virtue of "negotiated power" between state and industry, they generally work only at a later stage when industries are prepared to genuinely cooperate with government without serious conflicts of interest. It can be observed that the failure of asking industries to switch to E-T economy has increasingly become a serious headache, though bureaucrats have tried different ways to negotiate with industry. In some cases industries will promise to enter an agreement with government, but this does not necessarily mean that a positive outcome will eventuate. This phenomenon exists virtually in every country. For example, Japanese electric utilities had been fighting off the expensive wind technology by suppressing the transmission volume of wind energy in Tomamae, Hokkaido, of Japan. Similarly, only some Danish utilities that were locally owned, nonprofit making cooperatives behaved more cooperatively. Profiting-making utilities, however, had a more reserved attitude, begrudging mainly higher cost and tariff levied on immature industries, and wind energy was one such industry. On the whole, the Danish utilities did not behave very aggressively anyway. Up to 2001, only 14 percent of wind energy in Denmark was produced by "strong" utilities (the large majority of wind power was produced by "weak" individuals) of which many were for demonstration purpose, yielding to the government's executive orders, and not being caused by state-societal negotiations.

The overall picture above indicates that the negotiated public-private cooperation of GI did not play a role in the early E-T industry. Among the four types of GI, that is, "disciplined support," "public risk absorption," "private-sector governance," and "public-private alliances," none fits the polluting industry for E-T transformation. Although all of these have been successfully applied to the industrialization of NICs, they are not suitable for the special requirements of the E-T industry. Ample examples are available to demonstrate this.

First, the "disciplined support" way of promoting the E-Ts emphasis on performance or outcomes as preconditions to allocate state subsidies or protection. This method enables not only industrial motivation but also benefits by getting rid of rent-seeking behavior and better monitoring developmental progress. However, in the E-T industry, performance-based subsidization or protection is not workable simply because performance is out of the actor's control. The performance of early E-T businesses usually depends on many factors, for example, whether sunlight or wind force is strong, regular, and stable enough. In fact, due to this reason, the wind energy leaders, for example, Denmark and Germany, two of the leading wind energy countries in the world, have not applied such a conditional support to performance. Instead, they have been using fixed, guaranteed subsidies and a fixed wind selling price to every target actor; from 1988 onward, Denmark even gradually lowered (not rewarded) its support of 30 percent of investment and finally abandoned the remaining 10 percent in 1989 when the market of wind farms was found to be progressing. Also, successful E-T developing countries have not used penalties for persistent failure. The European packaging waste recycling policies are good examples. Although most industries could not increase their recycling rates on time, the European governments usually adopted a lenient attitude to extend deadlines and did not penalize their industries. Germany, for example, is one of the most lenient countries but achieved high packaging waste recycling rates. The government warned the industry that if the recycling rates did not reach the stipulated refilling quotas and the recycling targets (glass 90 percent, aluminum 90 percent and plastic 80 percent) by January 1995, a mandatory deposit system for all nonrefillable liquid containers would be established. Yet, as a result, the government gave a very long grace period (until January 2003) to its industries and retailers to retry again and again without penalizing them or installing the mandatory deposit system. This is totally different from the example quoted in Weiss's book that "according to JETRO, the government 'always refused to extend liberalization periods, with the specific intention of exhorting domestic manufacturers to pull out all the stops to meet the deadline' " (JETRO 1993: 139). In contrast, extension of deadline in E-T sectors is commonplace as most industries also counter threaten their governments with the probable economic harm to the country. So far as the packaging waste recycling sector was concerned, this moderate government attitude prevailed not only in Germany but also in other countries such as Sweden and Switzerland, resulting in high recycling rates at later stages, despite many of them initially facing "external" pressure to meet stringent targets or deadlines requested by, for example, the European Community.

The second major type of GI, "public risk absorption," is not suitable for the E-T industry either. In a conventional industry the Japanese or Taiwanese government has to absorb most of the risk to solicit cooperation, but the risk in the E-T sector is always far *too high* to be absorbed because it is about the public risk of *whole* industries. Weiss (1988: 75) quoted the Japanese machine tools example from the literature of Sarathy (1989: 142) that "MITI together with the Japan Development Bank organized a leasing company in 1980 that would enable domestic firms to lease robots under short-term arrangements and to return them without expense if dissatisfied with their performance." However, governments were unable to lease windmills or EFA cars to consumers or refund if the equipment was later found to be unsatisfactory. Actually, government can promote

E-Ts in many other ways but does not have the financial heft to absorb the risk of E-T project failure simply because it is too high to be assumed *only* by government. Actually, governments who have successfully developed E-Ts did not undertake the risk but only established laws and regulations so that utilities could purchase, for example, wind power at fixed selling prices. That is to say, risk and additional cost of renewable energy were not absorbed by governments but by utilities that then shifted part of the risk and cost evenly to consumers. Likewise, government cannot guarantee markets or absorb public risk for waste recyclers or environmentally friendly automakers. All waste recyclers, auto makers and auto buyers, have to take the plunge and share the additional cost. To emphasize the point once again, because the risk involved is too high (thanks to the staggering size of the pollution industry), government policies-even the most strategically planned Japanese ones-have refrained from guaranteeing markets to encourage the production or use, for example, of the then-successful gas-electric hybrid cars.

The third type of GI, "private-sector governance," is usually employed to save troubled or declining industrial sectors but not to encourage innovation. By forming cartels to control prices, setting production levels or import/export capacity (Weiss 1988: 76–78), such "private-interest governance" is however not workable in the E-T sector. If government delegates the power of general policymaking to industries in return for a promise to meet publicly defined criteria, the latter are of course in a much better position to generate a more benign business climate for themselves in order to improve their profitability. But such negotiated power does not guarantee that the industry will prioritize E-T development just because the E-T products usually belonged only to a very small portion of their normal businesses and profits. The utilities' indifference to the 1990 German "100 MW-Wind Program" in which the then Federal Ministry for Research and Technology entrusted industry associations with the governance of the program or the disappointing memory of the inadequate performance of the 1991 U.S. Advanced Battery Consortium (USABC) in which government gave the auto industry more self-reliance are two good cases to demonstrate the inapplicability of "private-sector governance" to E-T development. Unlike the Japanese circuit board industry or the Taiwanese textile industry that must satisfy every criteria set by the government for rewarding assistance or policymaking delegation since this is their *only* hope (Weiss 1998: 76–77), the electric utilities or automakers are more interested in their much more lucrative main business than in the government's criteria for upgrading the less lucrative or nonprofitable E-T production. Similarly, if governments delegate industries to manage their own waste recycling by using this type of GI to ask them to comply with recycling criteria, then we would be very doubtful of the effect because the main income sources of the industry do not come from the recycling business at all. These lessons taught us to seek another state-societal institutional setting to explain those rapidly growing E-T industries.

The final type of GI, "public-private innovation alliances," is a *genuine* public-private partnership but it very rarely materialized in the early E-T industry. The idea that government become a semientrepreneur—as seen, for example, in the Targeted Leading Product (TLP) program through which the Taiwanese government invested half of total development costs to support high-tech uptake in Taiwanese firms (Weiss 1998: 78)—has an advantage, in that it combines the strength of both "disciplined support" and "risk sharing." But, as explained earlier, these forms of GI are not suitable for the E-T industry. Although such public-private alliances for new technologies have been popular and successful in some NICs, it is observed, for instance, that the fast growing Japanese gas-electric hybrid auto technology was not realized by such a public-private alliance. It can be seen that the early clean fuel vehicle public-private R&D programs in Japan did not include the then-successful hybrid technology. In contrast, while the American PNGV program required the auto industry to work closely and brought about a very high government-industry spending ratio in the alliance, the result was however very disappointing. While public-private alliance may attract private participants from conventional industries to commit themselves to contributing a significant share of resources (Weiss 1988: 78), it fails to ask the monopolized power industries or the maverick auto industries to join forces or to share their capital or costs. Instead, for example, the Japanese government has refrained from investing so heavily in the public-private R&D programs working with the automaking firms, and ownership and/or technology was not transferred from unsuccessful firms to other firms either. The same happened in the wind energy sector. Governments never formed wind power technology partnerships with the largest equipment producers or utilities for technology diffusion. In stark contrast to the Taiwanese new Targeted Leading Product program, the Danish government-financed RD&D projects consisted of two major programs, that is, Ministry of Environment and Energy's Energy Research Program (EFP) and the Program for Development, Demonstration, and Information of Renewable Energy (UVE). However, both the EFP and UVE have not played a major role in propelling the Danish wind energy diffusion. Instead, the diffusion took place when investment aids and various supports were granted to small sized private owners who installed small turbines of 5 to 11 kW for their own use. The basic RD&D invested by the Danish Government working with the manufacturers only occurred at a later stage and the public support of these public-private innovation programs was limited to a small amount, that is, below US$2 million a year. The principal aim of these partnerships appeared to be the training and development of wind power operators rather than the commercialization of wind energy production. Søren Krohn, the Managing Director of the Danish Wind Industry Association commented on the Danish government's role in wind energy research in his report (2002: 3):

> An important difference between the Danish and other national governments' support for wind energy RD&D is that the Danish support has primarily been directed towards basic research, whereas other governments have tended to support wind turbine development. The difference is also remarkable when looking at the Ministry of Energy funding of energy research. The money in other areas of energy research tends to be spent much closer to the product development phase than is the case for wind energy.

In addition, the Danish public-private partnerships did not require firms to produce or sell the targeted product, let alone intellectual property rights, and thus they were not in the form of "public-private innovation alliances" described by Weiss. On the whole, governmental entrepreneurship and genuine public-private partnerships may not accommodate the nature of E-Ts and thus are seldom a workable public-private cooperative framework for a large scale national transformation such as the introduction of E-T industries.

Therefore, the above four forms of GI are part of the successful governance mode in conventional industries but not the one where there are still strong conflicts with E-T sectors. Apart from statism and GI, "state corporatism" and "social corporatism" are also classified under the same taxonomic group. According to the description in Weiss (1998: 37), they have different degrees of public-private relationships: "state corporatism" represents a "strong" state and "highly organized" industrial group to implement state-guided policy, but the industrial group is not "strong" enough (or sufficiently independent) to negotiate with the state (e.g., Park's era in South Korea and interwar Japan). "State corporatism applies where the state is 'strong' and the social group is highly organised, yet is more an instrument of public policy than a negotiating partner" (Weiss 1998: 37). Conversely, "social corporatism" refers to a "strong" industrial group but a "moderate" state, resulting in a higher level of self-governance by the industrial group (e.g., postwar Germany and Sweden). "Social corporatism applies where the state is moderately insulated and the social group is highly organised" (Weiss 1998: 37). Yet, both concepts of "social corporatism" and "state corporatism" are problematic for E-T development. In the state capacity model of industrial transformation, these concepts of public-private relationship provide reliable explanations for understanding the process of national development from one technology to another. However, on the condition that E-T development is regarded by many as having a zero-sum relationship between economy and environmental quality due to the practice of materialistic capitalism or other reasons (such as the omission of accounting the environmental capitals), the suffering industries would produce forces to counteract coercive state power. Thus, without a common interest, highly organized industries would use their economic and social strength to evade negotiation or seek to lobby the government, thus failing to achieve government goals. For these reasons, "social corporatism" with "strong" or "self-managed" industry would result in an E-T policy deadlock or abuse. Alternatively, "state corporatism" with "strong" state competence cannot be productive either when the "organized but submissive" industry is driven to a policy prodigality or abuse by a commanding government who is not familiar with the actual industrial operation and market. Further, E-T business is always placed at the center of pressure, especially in the situation of "corporate domination" under which business has an exclusive commercial control of an industry and prevents government from intervening (e.g., German or Japanese electric utilities, other monopolized public services, or dominant businesses). In any case, what is the best government-industry relationship for transplanting E-Ts to industries? I suggest a relationship between a "moderate" state and a "moderate" industrial group. In the following paragraphs, I will explain how such a "moderate" public-private relationship could break the policy deadlock and achieve desired E-T development goals?

On the one hand, after negotiations between government and industry last for a long time without making any progress, government would avert the vehement opposition of conventional industries by adopting "softer" or "moderate" regulations and other state efforts that provide industries with resource/financial assistances or grace period to adopt E-Ts. On the other hand, government would circumvent the most unwilling or defiant industries or actors and look for other target industries or actors who have a more "cooperative" or "moderate" attitude toward E-T adoption. "Moderate" or "obliging" E-T regulations devised by government officials are never so severe or unreasonable or inconsiderate. Governments are very careful to devise laws and regulations, trying to avoid unaffordable work loads or excessive financial pressures on industries or target actors. For example, the German government found it very difficult to motivate powerful monopolized utility companies to adopt the less-profit-making renewable energy even though government funds were offered and administration and monitoring tasks were delegated to industry experts in the 100-mw wind programs in 1990. A "top-down" mandate however incurred a lawsuit filed by the strong power companies in 1994 and a delegation to the industry unfortunately became discretion to refuse wind energy. The state eventually turned this impasse around by repositioning the policy target at those rural cooperatives that have much less opportunity costs to produce wind energy. At that moment general interest rates were low; the government guaranteed profitable prices for small wind energy producers; and new laws allowed farmers to purchase small pieces of land for wind projects. All of these had strongly attracted farmers and had rural communities rushing into wind power market.

That is to say, this process is *delicately balanced* by not exasperating and/or by relieving the distress of the conventional industry that has suffered from being asked to adopt considerable amount of unprofitable E-Ts. In the German wind energy case, the federal government comforted the most suffering utilities by amending the Electricity Feed Law (EFL) in April of 1998, asking each level of the utilities to equally share the additional cost by limiting the financial charges of the different utilities to 5 percent each (some of them were later even willing to absorb a higher charge of more than 5 percent). The engagement with renewable energy then spread through the country so quickly that the utilities did not want to lose market share and later changed their minds to absorb wind energy.

Under this *dynamic* and *balancing* institutional arrangement, the state strategically, carefully, and increasingly enlarges the E-T market without confrontation or causing serious suffering. E-T *permeation* occurs when social actors increasingly find E-T adoption affordable, inevitable, and/or obligatory, resolving the developmental predicaments of E-T, such as conflicts of interest, behavioral problems, and insufficient social or local participation. E-T diffusion takes place when policy applies to the most cooperative actors and simultaneously conciliates grumbling, disturbed industries that suffer from being asked to adopt expensive E-Ts; and in order to sustain the E-T diffusion, government maintains a delicate relationship between policy beneficiaries and potential sufferers when the former receives benefits without harming the latter. So this relationship is neither zero-sum nor win-win, but one that is somewhere in between. The E-T premium is largely absorbed by governments and consumers rather than *only* by industries in the E-T leading sectors.

The word "permeation" is to describe the process in which government regulations and other policies apply to cooperative market players first and then spread through them to other players. Yet, this particular permeation strategy has to be balanced by "pacifying" the potential sufferers/losers. In addition to the word "permeation," the new concept can thus be called "pacifying permeation" to denote a new form of state-industry relationship, not in terms of "strong" state versus "strong" industry, but in terms of the aforesaid "moderate" state versus "moderate" industry. By using the "pacifying permeation" tactic, the relationship between government and polluter-dominated industry becomes interdependent, as is the relationship between government and E-T developer. But the relationship between the polluter-dominated industry and E-T developer is not necessarily interdependent because E-T permeation and diffusion take place through a relationship in which the E-T developer derives benefits from the association while the polluter-dominated industry remains unharmed or unaffected. This specific triangular relationship is deliberately created by governments in E-T leading countries.

Therefore, when the social preferences for E-Ts are very diversified and unbalanced, they have to be treated separately by bureaucrats through a deliberate state-societal institutional setting that I call "pacifying permeation" as a new form of state-societal relationship as just explained. This is a supplemental concept of state-societal institutions in making the state, on the one hand, an active party to establishing new state-societal institutions for national transformation but also, on the other hand, making it a third party (a nonstake holder) regularly examining and adjusting the dynamic progress of institutions that link conventional industries and E-T actors. That is, tweaking institutions in the shadow of the state.

"Pacifying permeation" is actually a policy strategy. Its principal approaches are, firstly, to limit the size and extent of the newly institutionalized area until the deliberate institutional setting is strong enough to provide trustworthy rules and regulations that can then drive a more obedient or more willing group of social actors to act upon them; and, secondly, to incubate and protect this focused institutionalized area from being influenced or undermined by opponents or other social actors through the political structure. Usually this can be achieved by "pacifying" the opponents through "obliging" regulations and/or administrative means to relieve uncooperative actors from unrealistic requirements, financial hardships, and other onerous burdens. Such an efficacious institutional arrangement is a multistep approach, featuring a regular public-private interaction and a flexible regulatory approach. These two features do not connote different strategies but a dynamic institutional approach for E-T development, that is, an institutional arrangement subject to increase (and sometimes decrease) of regulations, state power, and target area in order to help E-T permeate the industry and market. To smoothly transform the industry into an E-T business, "pacifying permeation" will play an important role as if it were conducting a balancing act. For example, Denmark, one of the most successful wind energy development countries in the world, has seen state power that is equally moderate and concentrated right from the outset. Although executive orders were workable to ask the utilities to install windmill demonstration projects, reluctance was commonly found among the electric utilities. Søren Krohn, the former managing director of Danish Wind Industry Association, stated that there was "a more reserved attitude at the political level of (Danish) utility boards, basically resenting cost and tariff increases due to (costlier) renewables." To deal with this problem, unlike the wind energy laggard countries, the Danish state focused on those with an acquiescent attitude (the sectors with least resistance). As a result, the majority of early wind energy developers were farmers, rural individuals, and cooperatives. Meanwhile, in order to pacify the utilities that were obliged to pay wind energy developers expensive renewable energy feed tariffs (REFITs), the government arranged payment of 0.17 DKK/kWh from Energy Tax and 0.1 DKK/kWh from CO_2 Tax for the wind energy developers, providing some respite from the financial burden of the utilities. Further, the permeation effect continued when the utilities gained more confidence in a foreseeable wind business prospect. Having found that it was cheaper to generate their own wind power than to purchase it from individual wind producers, the utilities changed their mind, asking the government to grant them exclusive license for wind energy production. Their enthusiasm was shown when the first 750 MW wind farm ventured by Danish utilities was recently agreed to, without being asked to do so by an executive order.

To further demonstrate and generalize the phenomenon of the "pacifying permeation" institutions in which speedy E-T diffusion has taken place, I will construct analytical frameworks in the next chapter for a more systematical comparison and analysis of government policies in a number of E-T sectors and countries.

3

"Different Traits Design" Comparisons

3.1   Introduction

In order to examine the new institutional concepts, an analytical framework must enable us to compare and explain why some states are leaders in environmental-technology (E-T) diffusion and some are followers. This shall be achieved by demonstrating the impact of different institutions on E-T development and investigating whether their successful policies have employed the institutional arrangements conceptualized as "pacifying permeation."

The previous chapter examines the literature on institutions in relation to laws, regulations, and state-societal structures and relationships for national development, and it raises a set of criteria to determine which dynamic institutional arrangements led to the most efficient diffusion of E-Ts. Accordingly, the argument of needing to tweak institutions will be tested by choosing different institutions of each type for comparison. That is to say, it will investigate what their laws, regulations, and state-societal structures and relationships initially look like and how they are then changed. Similarly, the argument will also be tested by investigating whether the unsuccessful E-T policies lack the traits of "pacifying permeation."

This chapter will describe and explain the adoption of comparative case studies, sampling techniques and their restrictions, the procedural steps of evaluating evidence, and the sources of information used in each case study.

3.2   The Adoption of Comparative Case Studies

This book employs the case study approach as a way of discovering causal regularities and relationships between independent variables and policy outcomes (e.g., Van Evera 1997: 1–48; Gillham 2000: 9–14; John 2002: 222–223). Such a "variable-oriented" method has frequently been used in tandem with comparative studies since Przeworski and Teune (1970) exemplified the comparative technique of the Most Similar System Design. This technique is designed to find out whether or not different kinds of variables are in a genuine causal relationship by selecting and comparing cases that are similar in every respect except for the research focus variables. Cases that are as different as possible for testing which is the better for testing the causal independent variable (i.e., the Most Different System Design) are compared in this study (Prezeworski and Teune in Levi-Faur 2004: 20).

However, the increasingly complicated variations of regulatory reforms and state-societal structures and relationships in the E-T area suggest that a small number of case studies cannot completely cover all possibilities of variations. If the number of case studies is too small, then testing theories in comparative studies becomes doubtful. This is simply because such studies can easily be accused of "selection bias" or manipulation of cases to use only "minorities" or "anomalies" to answer research questions (Lieberson 1985).

To address the shortcomings of small-N comparison (usually less than three cases), many recommend the large-N or statistical method that usually has more than 50 cases (e.g., Lijphart 1975: 165). However, an excess of cases would undermine the strengths of contextual analysis, losing its essentially descriptive and detailed character and failing to get an in-depth understanding of the informal reality (new evidence) for theoretical implications. In addition, a large-N or statistical study would also create "assumption problems." Hall (2003: 381–382) states that "standard regression analysis and the comparative method understood in conventional terms provide strong bases for causal inferences only when the causal structures in the world to which they are applied conform to an exacting set of assumptions." Hall (2003: 384) also indicates that "some of the most prominent theories in comparative politics now understand the world in terms that do not conform to the assumptions required by standard regression analysis." In most research, assumptions are merely concessions or compromises when the events to be studied are so complex and/or the information is insufficient for a regression analysis. Hence, the number of cases studies in each of the later chapters is not large (unless they have very similar parameters in common), lest they compromise the strength of in-depth analysis and commit assumption problems. According to this basic principle, two cases each in chapters 4 and 6 and six cases in chapter 5 are thus chosen.

Furthermore, the theorists of both strategic interaction (e.g., Milner 1998) and path dependence (e.g., Pierson 2000) have shown that time (and history) plays an equally important role as other causal variables in the production of outcomes. They also suggest that immediate independent variables are not the only and strongest causes, instead, a wider and longer perspective should be considered. However, if the study of causal variables is contingent upon too broad a context, analysis would lose focus and homogeneity (Ragin 2000). Thus, a moderate breadth of events and length of time frame may provide the best outcomes. After having considered the requirements above, I confine the breadth of events to the three most significant and mature E-Ts—that is, wind energy, packaging waste recycling (PWR), and environmentally friendly auto technology (EFAT)—and the timeframe to the last thirty years. The selection of the cases is also based on the fact that they have strong characteristics of the three institutional components in question (discussed above). Further sampling techniques for these will be discussed later.

Given these premises, and in order to take full advantage of both small-N and large-N analyses without treating causal relationships as a timeless terrain, we must consider how best to arrange and compare the cases, as this will prove important for policy analysis and the answering of research questions. In this respect, there are a few comparative approaches, providing both the power of extensive observations for testing causal propositions and the virtue of inductive richness for causal inference. For example, Vogel (1996) compared different pairs of cases across two nations and two different times respectively in his first few chapters, performing a long series of small-N studies. Successively, he compared the same two countries with three new policy sectors one after the other, enhancing the degree of common regularities and irregularities. Furthermore, he added three new countries, contrasting their regulatory reforms in two of the last five policy sectors. Finally, he continued to compare the other remaining three policy sectors for the last three countries. When all was said and done, Vogel conducted a "texture" case study (that became an innovative medium-N case study technique on the whole) by progressively increasing the number of cases throughout the histories, countries, and policy sectors, producing a more convincing conclusion to his argument than conventional small-N or statistical case studies in regulatory analysis have done.

In addition, Levi-Faur (2002) theorized how to mix four widespread comparative approaches, namely the National Patterns Approach (NPA), the Policy Sector Approach (PSA), the International

Regime Approach (IRA), and the Temporal Patterns Approach (PTA) in order to maximize the explanatory power of medium-N comparative case studies. Levi-Faur (2002: 6) labeled this technique "complex research design" and pointed out that "the important terrain of medium-N research (more than two and less than circa 100 cases) is *terra incognita* for many." In addition, Levi-Faur also pointed out a possible method of systemic examination by looking at four different combinations of cross-sectoral and cross variations in order that both a "panoramic snapshot" and a "holistic picture" may be acquired.

Such medium-N case study techniques, which enable both qualitative and quantitative investigation of argument, have provided us with a fundamental framework design for broadening dimensions in order to scrutinize evidence. By doing so, I examine cases systematically across nations, sectors, international regimes, and time, as per what Vogel conducted and Levi-Faur proposed. But, for a stronger analysis, I modify them into a "different traits design" to test the propositions, that is, in my analysis different policy sectors in a single country might have different policy traits. For example, the United States has very different policy traits across its energy, PWR, and EFA industries, respectively employing deregulation, regulation, and government-industry cooperation. They are also related to the institutional literature in question, so it is very useful and proper to scrutinize these different policy traits thoroughly. Under the "different traits design," similar institutional traits of cases are *neither repetitively demonstrated nor mechanically compared* from every possible perspective of countries and sectors. Despite countries not being randomly chosen, they are *not* purposely selected to fit the author's argument. In fact, the second group of case studies is chosen only because they have similar institutional traits (i.e., heavy regulations) but different policy outcomes, challenging the first case study group's results. Thus, while the first case studies have proved the importance of regulations, the second group of case studies helps argue that not only regulations but state-societal structures also are causal factors. In turn, the third group of case studies between two conflicting policy outcomes under similar government-industry cooperation may help clarify why regulations and state-societal structures are still insufficient for achieving desired policy outcomes.

The case studies will start with a comparison between two political economies with conflicting institutional traits. Based on the "most different system design" and the "most similar system design" as discussed above, wind energy development in the United Kingdom and Germany are compared because this particular sector of the former is famous for its *deregulation* and the corresponding sector of the latter is heavily characterized by its *regulation-based* industrial policy, though generally they have other similar traits such as their common European energy legislation. I then compare cases with a *new sector trait*, that is, cases no longer regulatory but having state-societal structures. Thus, PWR developments in the United States, the United Kingdom, Australia, Germany, Sweden, and Switzerland (which have different state-societal structures but a similar trait, that is, their highly regulated PWR markets are centered on container deposit legislation) are compared. And then, EFAT developments in the United States and Japan (which have different degrees of state-societal relationships, that is, "the rule of law" versus "the rule of bureaucracy," but similar government-industry partnership programs) are also compared.

By using this "different traits design" to test propositions, the final texture of the comparative medium-N case "fabric" is "knitted" to reflect step by step—from the generalized to the specific—different forms of institutions. This also follows from the spirit of theory-building across countries and sectors of different policy traits and the analysis of negative evidence through "large" to "small" traits. It means that, on the one hand, the combination of case studies is strategically designed for filtering common institutional characteristics (forming a common institutional concept for E-T development) from different spectra of a group of case studies. On the other hand, settings of subsequent groups of case studies for comparison are totally dependent upon the traits of their policy sectors.

The first group of case studies should be *larger* and *wider* enough to establish "coarse" explanations to the research question: what do the regulations look like in those successful E-T policies? Although the bulk of evidence in the first group of case studies may have answered the central questions, there may still be negative evidence that does not conform to the initial hypotheses. For example, despite having discovered that regulation rather than deregulation plays a more important role in wind energy diffusion, it does not imply that regulation is the only independent variable if conflicting policy outcomes are seen in other policy sectors that are heavily regulated. Thus the hypothesis would have to be refined to explain why negative evidence exists and to produce a more plausible hypothesis. The common institutional characteristics of countries that have successfully adopted E-Ts in one E-T policy sector may not necessarily be identical to the institutional characteristics of countries with successful E-T adoption in other E-T policy sectors. For example, the packaging waste recycling cases may demonstrate that not only regulation but state-societal structures also matter. Therefore, a second or even a third group of case studies for a closer study on common institutional features would be required until no contradictory evidence is found.

3.3   SAMPLING TECHNIQUES AND THEIR RESTRICTIONS

Design for the Whole Case Study

The design of this case study method is a specially arranged exhibition of comparative cases at different angles and times, in a logical, inter-active, and investigative manner, to systematically confirm the hypotheses and answer the research questions. In other words, some distinct E-T policy sectors were consecutively compared and analyzed along the thread of hypotheses in terms of both "breadth" and "depth." In doing so, three sampling principles were used.

First, the case studies were designed to go along with *the thread of hypotheses.* Following the findings of the wind energy comparative case study that a highly regulated market and intensive public-private cooperative programs are characteristics that it shares with a "leading" Germany over a "sluggish" United Kingdom, it is worth investigating why there are still different policy outcomes in E-T policy sectors in which some political economies have highly regulated markets or intensively connected public-private partnership programs. Thus, the highly regulated market of the PWR industry and the intensively cooperative public-private partnerships of the EFA industry were selected. Accordingly, the first comparative case study is to analyze why liberal market economies (LMEs) and coordinated market economies (CMEs) have had different E-T policy outcomes over time, and to study deregulatory and regulatory institutional environ-ments that have had to comply with EU liberalization directives for electricity markets. The second comparative case study is to investi-gate the institutional characteristics with a special focus on the state-societal structures between LMEs and CMEs in a *highly regulated* E-T policy area. The third comparative case study is to intensively examine the institutional characteristics with a special focus on the state-societal relationships between LMEs and CMEs in a public-private partner-ship policy sector.

Second, the balance between *breadth* and *depth* was also considered in the design. Each group contained a country in the wind energy policy sector, comparing and measuring their regulations and state powers in terms of breadth. Next, the breadth of the samples was widened by extending to other sectors and some more countries. Thus, LMEs and CMEs were tested again by choosing three countries from each group in the highly regulated PWR policy sector. Since both the first and the second comparative case study grids had a considerable number of samples, generalization of data could be readily carried out. However, generalized data can overlook some intervening and/or condition variables that are required to answer the second stratum of research questions about the significance of the degree and extent of state power in the relationship between institutions and E-T development. Thus, a detailed comparative study on state capacity and policy tactics logically led to a third case study on the EFAT policy sector, with a detailed comparative focus on two different public-private partnerships. Each of these three E-T policy sectors, an LME group and a CME group, consisted of data collected from two to six advanced industrial countries over the last few decades. On the whole, the combination of these three comparative case study grids has covered not only the most important and mature E-Ts but also the principles for finding a balance between breadth and depth in social science methodology (Ragin 2000, Levi-Faur 2002/2004). And more significantly, the degree of breadth and depth of the cases herein explored were arranged according to the thread of the hypotheses, and not mechanically according to every possible approach.

Third, further samples were selected based on *negative evidence* found in preceding case studies. Although the wind energy cases demonstrated some positive evidence that the CME country, (i.e., Germany, after the mid-1990s) have overwhelmingly outperformed the other LME country, that is, the United Kingdom, over the same period, the wind energy cases otherwise have no sufficient evidence to support any argument. Some negative evidence shows that Germany before the mid-1990s did not achieve speedy wind energy diffusion, and the United Kingdom experienced a sudden growth of its wind turbine installation in the early 2000s. Thus, I need to further study their institutions over these periods and also other E-T sectors characterized by heavy regulation and government involvement to look for both positive and negative evidences. For this purpose, the highly regulated PWR sector and public-private cooperative EFAT sector are suitable comparative case groups to reveal more facts both generally and specifically.

In general, the countries were not purposely selected to fit the author's arguments, but were chosen by the above-mentioned three sampling principles. The whole process can be compared to detective work: logic and clue finding; thus it may be labeled as "different traits design" medium-N comparative case studies. This design has the advantages of large- and small-N case studies without compromising the strength of generalization and specification but leading to more accurate and convincing research answers (see the illustration in chart 3.1).

The countries were selected from the most typical LMEs and CMEs among OECD members; so that more comparable aspects and

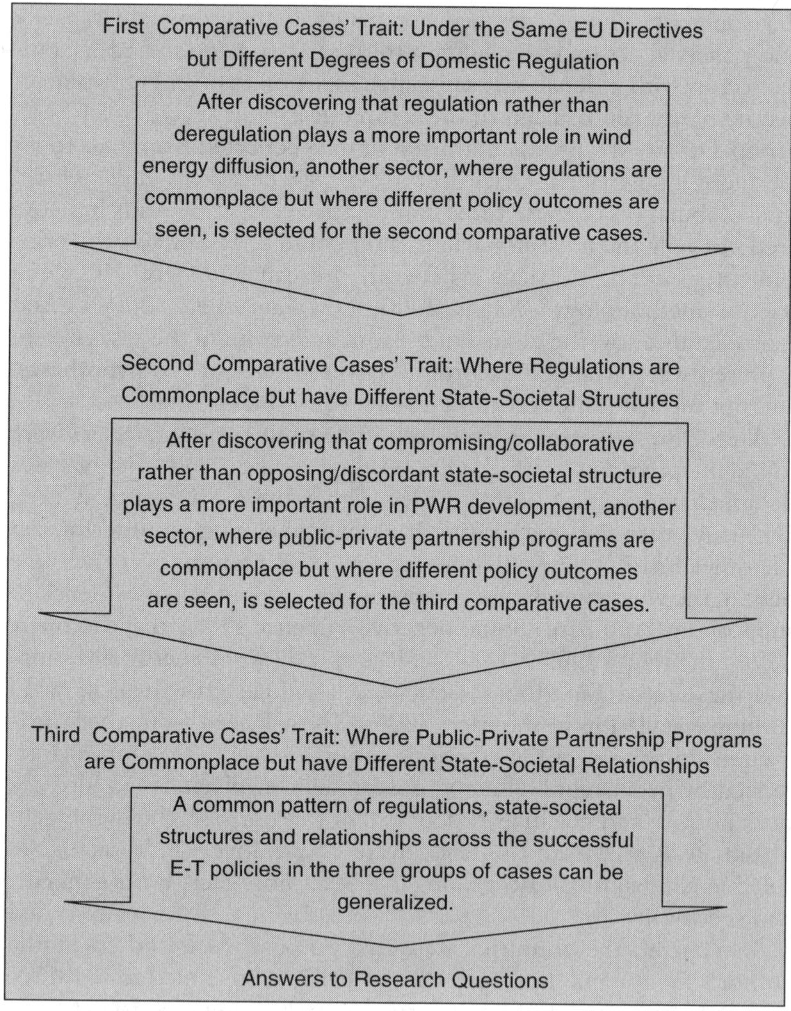

Chart 3.1   Sector-Trait Case Study Design.

reliable data were guaranteed. The LME samples were selected mostly from the United States, the United Kingdom, and Australia, while the CME samples were from West Europe and Japan. These selections are based on the descriptions and definitions of LMEs and CMEs in the literature by Hall and Soskice (2001: 8–68). Other works in the same publication edited by Hall and Soskice have revealed that these two markedly contrasting institutional arrangements have led to different developmental outcomes in many industries. Given the explanation in the introductory chapter that E-Ts have a very special nature, these bipartite institutional structures are herein retested to investigate whether environmental technology development patterns and performance are also different under differing institutional arrangements and whether further new knowledge of state capacity can be induced. In addition to the design for the whole case study, each individual comparative case study has its own sampling technique for conducting comparisons as discussed hereunder.

Sampling Technique for the First Comparative Case Study

Both LME and CME groups were firstly tested by the wind energy sector, a mature industry having some twenty years' track records. An LME group and a CME group, each consisting of one country, were compared. These comparative case studies focused on wind power because it is one of the most mature environmental technologies and is now used as an energy source in many developed countries. This policy sector may provide official and comprehensive information over the course of its development. Evidence and data were collected from the United Kingdom and Germany, as the former has had typical traditional LMEs and the latter is a well-recognized representative of CMEs. Each has an abundance of areas conducive to wind harvesting and has been a pioneer of wind power development in the last two decades. In addition, they are members of the OECD and many other international organizations from which reliable data and sufficient environmental technology records can be obtained. Furthermore, energy production is one of the most essential and influential requirements of both the environment and human life. These two countries have remarkably different economic structures, especially in terms of the degree of institutional regulation in their energy markets. Thus, this comparative case study naturally focused on the relationship between institutional regulations and wind power production and diffusion. Comparative studies between these two economic groups were conducted to investigate whether they had different wind energy policy patterns and outcomes and to indicate whether they have been successful over the last two decades.

The wind energy policies under examination were implemented in two stages. The first stage took place in the 1980s and early 1990s: a period of new institutional implementation for promoting wind power production in response to the need for alternative and clean energy sources. The second stage from the mid-1990s to the early 2000s was a period of further institutional improvement aiming at a fully functioning wind development system. Attention will be paid especially to this period because since the late 1990s wind power has been the quickest-developing new energy source globally, achieving an average growth of 30 percent a year.[1]

Sampling Technique for the Second Comparative Case Study

After conducting the wind power comparative case study, the PWR comparative case study followed, placing Anglo-Saxon LMEs and European CMEs on the test-bed. This sample selection was based on the grounds that they were at the frontier of making PWR laws, having at least twenty years of policy implementation with considerable and noteworthy track records. Furthermore, they are famous for their drink bottle waste management. The beverage container as one of the most significant and popular packaging types was brought into focus to facilitate the search for actual institutional factors in the making of PWR "leaders" and "laggards."

It is very hard to select places with very similar physical particulars. Although the size and population density of these six countries are quite different, the two factors do not really account for the difference in performance of their PWR programs because, for example, Germany and Sweden are the densest and sparsest respectively (237 and 22 inhabitants/km^2) among the six cases, but both acquired high recycling rates. Similarly, the United Kingdom and Australia are respectively dense and sparse countries (284 and 2 inhabitants/km^2 respectively) but both have relatively lower recycling rates than Germany and Sweden. Furthermore, landfill ban or subsidization also do not influence recycling performances in these six cases since they are more or less in the same "boat," employing similar strategies to control the volume of landfill and trying to switch to recycling.

For this study of different institutions, commonalities and variations of state structures in terms of legislative and executive powers and their relationships within which PWR laws and programs are designed, made, and implemented, and government-industry relationships for PWR development between "leading" and "laggard" countries are examined. This comparative study comprises two state-societal structural dimensions. First, vertically, do differences in legislative tradition and government structure, particularly in the relationship between the central and local governments, give rise to diverging national and local enactments and implementations of PWR laws and regulations? Second, horizontally, in the process of PWR law and program making and implementation, do the "discord" legislature and "majoritarian" executive produce constraints that limit the national development of the PWR industry? Alternatively, do the "compromising" legislature and "coalition" executive exert enabling effects that boost the PWR rate? Overall, do their institutions in terms of state-societal structure in which national and local governments maneuver dictate what they can enact and how they can work with societal groups, thus leading to divergent results?

Sampling Technique for the Third Comparative Case Study

The third case study on EFAT development was a detailed comparison focusing on the institutional difference, especially the state-societal relationship, between just two countries. The United States and Japan, rather than the United Kingdom and Germany, were selected. This arrangement for comparison is due not only to the former two countries being much larger in production size than the latter two,[2] but also to the possibility of obtaining a more comparable specification in the latter two. (This is despite the fact that Germany has already reached a 3L per 100-km target, equivalent to 80 mpg since 1997, and, its deficiency of vehicle size requirements has outstripped the former two countries.)[3] These two cases (the United States and Japan) were compared also because they already have conspicuously different outcomes. The government-industry cooperative programs in the United States were declared dead and targeted products were never produced, while those in Japan resulted in substantial sales in the market.

The first section of this case study chapter examines the context for the emergence of the American Partnership for a New Generation of Vehicles (PNGV) and the formation of its institutional arrangement, explaining why the American state-societal relationship did not promote the PNGV. This was followed by a detailed study of the corresponding case of Japan, demonstrating how active—but flexible—bureaucracies facilitated an effective state-societal relationship, stimulating the innovation and accelerating the development of EFAT.

Restrictions of the Methodology

These country samples were deliberately chosen from the most typical and developed LMEs and CMEs in OECD countries on a basis of the "most similar system design" and the "most different system design" principles. Hence, the primary concern in the case selection is to test whether the most distinct political economic structures will have different E-T policy patterns and outcomes in response to similar E-T development challenges and whether the outcomes are attributed to their institutional arrangements and functions. However, it is necessary to point out that the E-T policy outcomes, under any single group, LMEs or CMEs, are not necessarily identical.

Firstly, because some unique domestic or international factors may have influenced the timing or progress of the policy implementation. Yet, these delayed policy outcomes should not be regarded as usual phenomena and will not disprove the generalized explanation on a long-term basis as long as some reasonable explanation (e.g., renewable energy agenda in energy-abundant countries such as Australia would be less pressing) can be obtained to justify the delay. In any case, countries significantly impacted by noninstitutional factors are excluded from the samples to avoid confusion.

Secondly, the degree of liberalization or coordination differs from country to country. Even though the selected countries are overall archetypal LMEs or CMEs, the governments might not have exerted the same type or level of institutional arrangement and state power on their wind energy sectors. As a result, the policy outcomes of the countries might deviate from each other, though they are of the same case study group. Yet, the study on the policy outcomes concentrated on and evaluated not only the volume but also the pattern of the energy diffusion. The investigation covered a long span of time (over the last two to three decades) and focused on the impact of the various degrees and extents of regulation, state-societal structure and relationship on the policy outcomes.

Thirdly, most of the above questions could be answered by quantitative and/or qualitative analyzes but it was difficult to measure the overall performance of the E-T policies. An E-T policy that succeeded in achieving its objective was only an apparent result. The E-T that developed with a net growth rate (minus the subsidies and support) was more highly regarded as a satisfactory success. However, it was difficult to measure the cost of subsidies to RD&D that was a public asset. Nevertheless, the basic criteria for evaluating whether the E-T policies achieved their targets is the flow of technological change (Aghion and Howitt 1998: chapter 5). The overall rate of technological change is treated as an indicator of substantial magnitude for green accounting (Weitzman 1997). Alternatively, the indirect measurement of E-T improvement can be a practical way to present the outcome of E-T policies. For example, the extent of resource-saving, technological progress may determine the onset of constraints arising from reliance on nonrenewable resources in production (Nordhaus 1995). However, it was impossible to measure the full value of technological change through standard national accounting practices because the value of the resources that the government invested in RD&D for E-T actors was also a public asset. Nevertheless, a gross outlay on E-T investment (e.g., expenditure of money on E-T programs) and/or a general observation on E-T improvement (mostly in terms of the rate of diffusion) were carried out for reference and comparison.

In any case, this book is to investigate, analyze, and explain the relationship between institutional features and E-T diffusion. Further research and assessment on whether their E-T policies are successful or not, or whether CMEs have the absolute ascendancy over LMEs or vice versa, or whether a combination of the two is the need of the hour in the E-T territory is an alternative aim of this book and of course a more precise and overall measurement should be left to large-scale, further endeavors. Nevertheless, as a conclusion to the research findings, having digested the assorted methods of analyzing public policies, I have analyzed the performance of E-T development through an institutional explanation vis-à-vis other conventional arguments.

4[1]

Varieties of Regulatory State: Wind Power Diffusion in the United Kingdom and Germany

4.1   Introduction

Recently, new governance modes have been the subject of wide discussion in regulatory state literature (Moran 2001a&b/2002; Cook, Kirkpatrick, Minogue, and Parker eds. 2004; Jordana and Levi-Faur eds. 2004/2005). Writers refer to "the rise of the post-regulatory state" (Scott 2004) and "the making of a new regulatory order" (Jordana and Levi-Faur 2005). Typical is the view of Levi-Faur (2005: 5) that "regulation is a necessary condition for the functioning of the market . . . regulation is helping to legitimize markets and facilitate transactions by enhancing trust." This rise of "pro-competitive" and/or "promarket" regulation is one of the defining features of these models of a new regulatory order. Substitution of state discretionary power with market players' decisions is not the aim, however. The state's regulatory role is changing from "setting down rules and powers" to "meta-regulation," that is, a steering role that includes "legal underpinning for indirect control over internal normative systems" where, nevertheless, "ends are ultimately set and determined by the sovereign state" (Scott 2004: 167–168). Yet, all these arguments for a more light-handed regulation and a steering role of the state appear insufficient when implementation is further closely scrutinized. What do the national institutions and their regulations exactly look like during implementation? Does it refer to a general trend of regulatory convergence across liberal and coordinated market economies? If so, does this suggest a more effective policy approach towards E-T development?

A comparative case study into the recent regulatory reforms of the wind energy markets of the deregulated United Kingdom and (self-)regulated Germany is presented here in order to throw light on these latest institutional developments. The emergence of wind power as an alternative energy source has increasingly depended on a more diversified and popular renewable energy (supply and distribution) sector. Both national governments give high priority on keeping one step ahead of the technology revolution and regulatory innovation through appropriate policy settings. Their principal objective is similar, even if some of their policies (discussed below) are different: each country manages to comply with the increasingly stringent cross-border European directives and, of course, aspires to continue to being an "energy hub" (while the United Kingdom has long been a regional hub of oil and gas, Germany an exporter of coal), though renewable energy is presently far from being treated as an export product. In achieving this, they have two main concerns: first, how to ensure that renewable energy operators pursue not only profitability (or at least commercial viability) but also diffusion of the new energy to the public, and second, how to ensure (in the processes of British deregulation and German (self-)regulation) that they engage in capital- and technology-intensive long-term investment for the sake of technological and regulatory innovation, and for complying with the regional requirements.

European Wind Energy Market: The Challenge in the 1990s

Since 1990, the European Commission has initiated a single and competitive energy market for the region. One of the most important European regulations has been Directive 96/92/EEC that required most of its state members to incorporate corresponding regulatory changes for creating an open and competitive electricity market by February 20, 1999. As a result of this policy directive it was wind energy, one of the most promising but still very expensive new energy sources of the early 1990s, that faced severe "ordeals" in the two most policy-conflicting member states: the United Kingdom and Germany.[2] Further to the (post-)regulatory state literature above, three spontaneous questions arise:

First, has the market force strengthened by the European regulations for increasing regional liberalization process successfully developed wind energy in the "deregulated" British electricity market?

Second, has the development progress of the embryonic wind power industry been seriously stymied by the European energy-market liberalization requirements in the protected (or self-regulated) German electricity industry?

Third, is market-oriented "deregulation" or industry-led "self-regulation" a more effective policy option for national wind energy development in place of the European light-handed and steering (or nonlegally binding) regulatory backdrop, and if so, why?

Institutions refer to formal and informal rules, such as laws, regulations, market and organizational norms that actors generally follow (North 1990: 3; Hall and Soskice 2001: 9). This institutional approach has been used to explain recent national "regulatory" differences between the United Kingdom and Germany. Fioretos (2001) used it to explain why national preferences of the United Kingdom and Germany could exist in the era of European integration, contending that their individual political and economic structures, rather than the European regulations, shape the regionalization process. Vitols (2001) examined how corporations interacted with their national institutions in these two member-states, arguing that, despite the more open market, their specialization in different types of innovative business is increasing rather than decreasing. Wood (2001) explained that the British and German labor market policy differences were an outcome of their different patterns of government-business institutional relationship. Casper and Kettler (2001) attributed the fact that Germany has experienced more rapid biotechnological growth than the United Kingdom to the "hybridization" of business strategies in the national institutional framework. These analyses of institutional impact have extended the "varieties of capitalism" and "regulatory state" theoretical perspectives, but the impact of institutional and regulatory configurations on renewable energy uptake is still unexplored, especially between the two extremes of British "deregulatory" and German "(self-)regulatory" state power that are subject to the same challenge and starting point.

My argument is not that these two wind power development approaches can compete with increasing electricity liberalization and regionalization, but rather that "obliging" state powers that are compatible with their respective institutional structures enabling to tweak their own institutions are vital to the outcome of wind energy development by harnessing the state's capacity to smoothen out technology diffusion. As explained in chapters 1 and 2, the word "obliging" in this book refers to the willingness of government to be helpful without causing panic among regulated actors, particularly the utilities in this case study, so that they can tolerate stringent regulatory arrangements (e.g., the adoption of expensive wind energy) without fighting back. In other words, the word "obliging" is used to describe a "non-confrontational nature" of regulation used at a time when government has to change its attitude toward a defiant industry. However, the employment of "obliging" regulations does not mean that government loses its power. Rather, it is only one of the tactics of the government in the whole transformation process to achieve desired development goals.

This chapter analyzes the differing effectiveness (in terms of the volume of wind energy uptake) of wind power policies between the two countries by using the institutional approach to compare government laws and regulations, and the triangular relationships of wind energy developers, utilities, and their governments, along with the two different national wind energy market regulatory structures.

I will therefore examine the dilemma of market forces and state involvement in wind energy development, the methodology of this analysis, and two comparative case studies. The British case study will be scrutinized first. The empirical evidence obtained by using the institutional approach explains why the market-oriented tendering system in the United Kingdom over the 1990s did not work and led to only a paltry amount of wind energy production. The liberalization of the British electricity industry throughout the 1990s resulted in a sharp increase in the use of cheaper gas power, not expensive and troublesome wind power. The evidence also identifies the solutions to wind power developmental problems implemented by the government in the early 2000s. The next section of the chapter analyzes the case of Germany. It illustrates why its old form of state power did not adapt to the new era and how the government modified this old form of coordination, finally resolving the problems and making much faster progress in wind power development. The concluding section explores the implications of the case studies for clarifying and confirming the advantages of state involvement in the two different domestic institutional structures.

4.2   THE DILEMMA OF MARKET FORCES AND STATE INVOLVEMENT

The Emergence of Wind Energy

For some decades now quite a number of governments have committed themselves to greatly decreasing their greenhouse gas emissions and to developing alternative energy sources to fossil fuel. Providing a nonpolluting and renewable source of energy, wind power has been seen as a promising environmental technology with the potential to solve many of the environmental and energy shortage problems. Over the past two decades, wind power has been one of the world's rapidly expanding new energy sources (chart 4.1). Europe has been the top market for wind energy worldwide. According to the European Wind Energy Association (EWEA), wind power accumulative capacity in the EU increased from 439 MW in 1990 to 23,058 MW in 2002 (chart 4.2). Today, 75 percent of global wind power capacity (showing 35

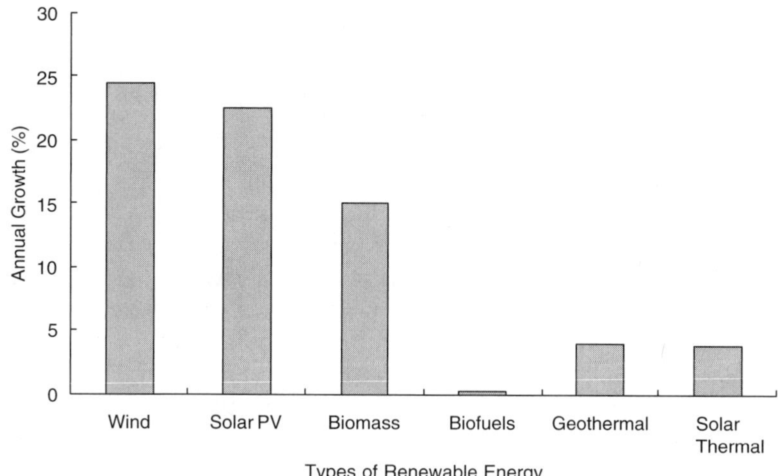

Chart 4.1 Growths of Wind and Other Renewable Energies.
Source: Energy Research Centre of the Netherlands ECN.

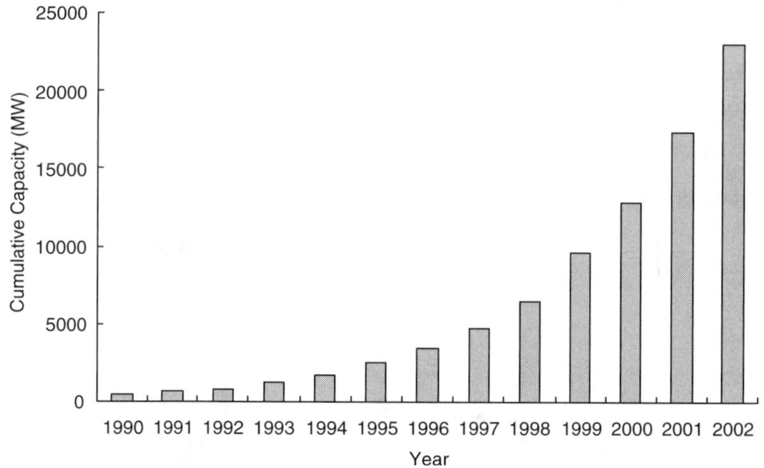

Chart 4.2 Wind Power Cumulative Capacity in EU, 1990–2002.
Source: EWEA.

percent average growth rate) is installed in Europe, and 85 percent of the global wind market is supplied by European manufacturers.[3]

The Problems of Wind Energy

Wind energy development problems, however, have obstructed wind technology development progress. High investment costs, frequent conflicts of interest, and market failures are contributing factors. In general, the generating cost of wind power has been higher than that of conventional energy in the 1990s, even though the gap has become smaller in the early 2000s (chart 4.3). Moreover, with traditional electricity infrastructures tending to favor fossil-fuel energy sources, wind power has been restricted to small, localized production modes. For these reasons increased future government involvement via subsidization, R&D support, conflict settlement, and market fixing would seem inevitable to counter past systemic failure. However, the dilemma of market forces and state involvement in wind energy development has a political dimension, making the energy market scenario seem perplexing at times. The following paragraphs will examine the conflicting arguments for and against government involvement in wind energy development.

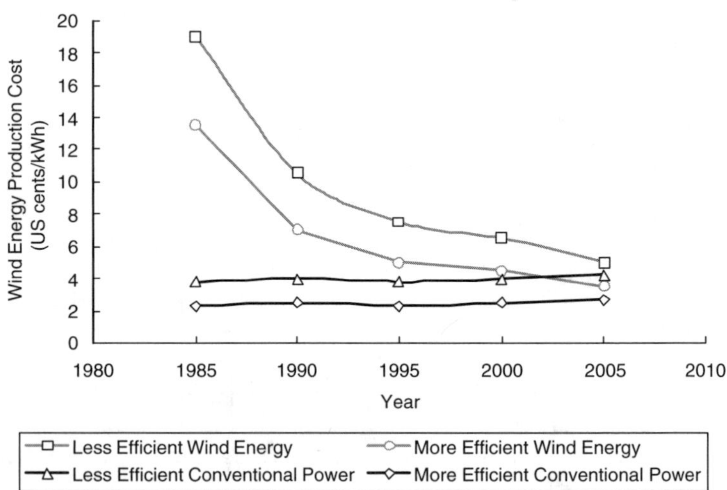

Chart 4.3   Costs between Wind Energy and Conventional Power.

Source: EWEA via ECN.

Arguments Comparing "Deregulation" and "Regulation"

The extent and degree of state involvement in wind power development has for long been a difficult and complex policy issue. On the one hand, the prevalent view of "deregulation" and "free-market" economists is that "regulation" is a constraining force, and that resorting to state power to support costly industry and technology contradicts previously successful free-market economic theories. Countries such as the United Kingdom in the 1990s that are of this opinion emphasize the importance of market forces and price signals on economic development. Their economic policies usually favor market liberalization and entrepreneurship. This explains why over the last few decades the electric power industry in many countries under the influence of economic neoliberalism has undergone "deregulatory" structural reforms, with wind power being introduced amid these "deregulation" episodes. By looking at the first decade of neoliberal reforms, quite a few economists have held the opinion that "regulation" reduces economic efficiency (e.g., Peltzman 1989; Noll 1989; Duch 1991; Niskanen 1993).

On the other hand, wind has been a very expensive source of energy costing in the early 1990s approximately four times fossil fuel energy. Despite technological improvements continuously reducing wind energy production cost, wind has still not replaced conventional energy sources as a commercially viable alternative. If governments do not actively support vulnerable wind energy industries, market forces will have a depressing effect on this noncompetitive technology and on the commitment to having clear air and energy security. The European wind energy market would have seen fierce price competitions if the EU requirement that electricity market liberalization be implemented by 1998 had been imposed, even though the other external institutions such as the 1997 Kyoto Protocol and the 2001 Marrakesh Accords did not require developed countries to implement emissions targets until the early 2000s.[4]

This latter argument supports the notion that nation-states should have greater involvement in the technology coordination process, often involving cooperative government-business relations and domestic regulation. Although few scholars held different (and supportive) views on institutional rules and regulations (May 1985; Hills 1986; Moran 1991), the debate in the early days lacked balance. Along with more and more evidence from the 1990s market, better organized empirical studies on the positive impact of institutional rules and regulations testify that "regulation" has not really been abandoned by governments in advanced industrial countries (e.g., Vogel 1996; Busch 2002; Grossman 2003).

The Impact of EU Electricity Liberalization on Domestic "Deregulation" and "Self-Regulation"

A single and competitive energy market policy has been in place in Europe since 1990. The directive on the Transit of Electricity through Transmission Grids (90/547/EEC) and the directive on the Improvement of the Transparency of Gas and Electricity Prices (90/377/EEC) were introduced by the European Economic Community (EEC) in 1990 as an initial common European energy legislation designed to meet this policy objective.[5] Notably since the Treaty of Maastricht and the European Union (EU) were established in 1992, the member states have accelerated the push to integrate their energy policies. In this context, further common rules for the internal market in electricity were introduced (directives 96/92/EEC and 98/30/EEC in 1996 and 1998 respectively), requesting the member states to dismantle their electricity monopoly and to follow a market liberalization path.[6] It implied that competition for a lower electricity price would be in place soon. Although the member states were not requested to fully liberalize and implement these directives immediately (but not later than February 1999 and August 2000 respectively to incorporate the EU provisions into domestic legislation),[7] the impact jolted many national electricity industries throughout the 1990s into making major organizational and institutional changes in anticipation of the perceived challenges to come.

In the aftermath of the EU electricity liberalization, the volumes of diffusion in terms of megawatt installation of wind energy in the United Kingdom and Germany over the last decade were very different (charts 4.4, 4.5, and 4.6). For example, according to the EWEA, while the accumulated volume of German wind energy growth was 2,018 MW in 1997, that of the United Kingdom was only 319 MW in the same year. The difference was even more remarkable in 2002. What caused this great difference? It is worth investigating the effects of market-oriented "deregulation" or industry-led "self-regulation" by studying the interactions between the governments, the utilities, and the wind energy market players.

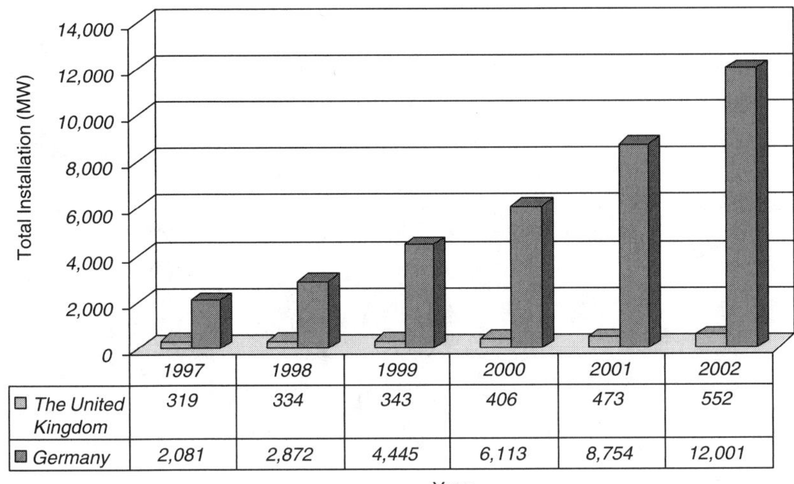

	1997	1998	1999	2000	2001	2002
The United Kingdom	319	334	343	406	473	552
Germany	2,081	2,872	4,445	6,113	8,754	12,001

Chart 4.4 Wind Power Total Installation (MW).
Source: EWEA, GWEA, GWEI, IEA.

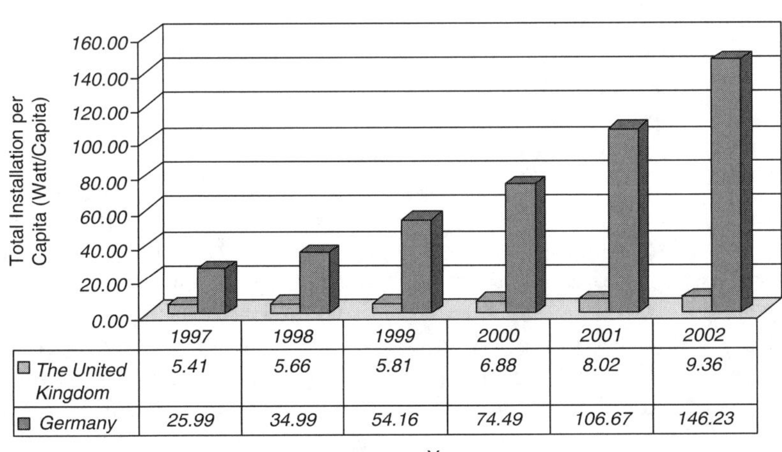

	1997	1998	1999	2000	2001	2002
The United Kingdom	5.41	5.66	5.81	6.88	8.02	9.36
Germany	25.99	34.99	54.16	74.49	106.67	146.23

Chart 4.5 Wind Power Total Installation per Capita (Watt/Capita).
Source: EWEA, GWEA, GWEI, IEA.

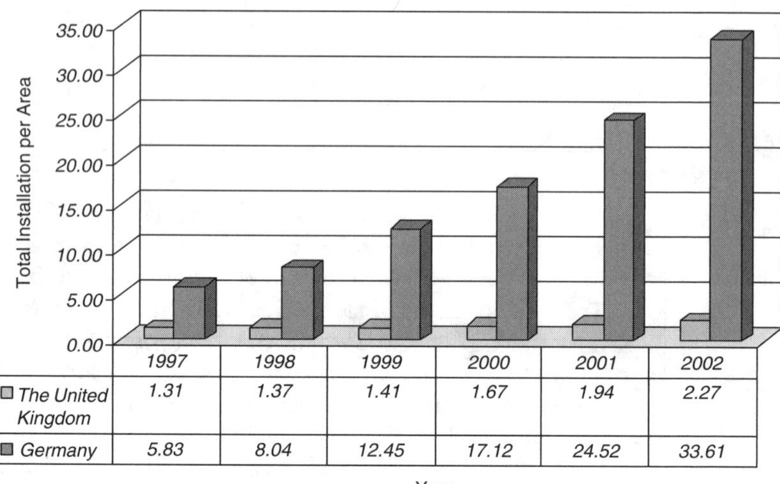

	1997	1998	1999	2000	2001	2002
The United Kingdom	1.31	1.37	1.41	1.67	1.94	2.27
Germany	5.83	8.04	12.45	17.12	24.52	33.61

Year

Chart 4.6 Wind Power Total Installation per Area (kW/kM^2).
Source: EWEA, GWEA, GWEI, IEA.

4.3 WIND POWER CASE STUDY METHOD

As mentioned in chapter 3, case studies under the "different traits design" will start with a comparison between two political economies with conflicting institutional traits. Evidence and data were collected from the United Kingdom and Germany, as the former is generally a typical and traditional liberal market economy (LME) and the latter is a well-recognized and representative coordinated market economy (CME) according to the "Varieties of Capitalism" comparative institutional approach (Hall and Soskice 2001: 8–38). These comparative case studies focused on wind power not only because it is one of the most mature environmental technologies providing official and comprehensive information over the course of development, but also because it creates a policy dilemma of market forces and state involvement. The United Kingdom and Germany, each has a considerable size of windy area (the former is one of the windiest places in Europe) and has been a pioneer of wind power development over the last two decades. These two nations were deliberately chosen from the EU countries on the basis of the "most similar system design" (MSSD) and the "most different system design" (MDSD) principles. That is, countries that have most similar energy targets, sizes, locations, and EU requirements but obviously most different institutional structures are compared. Hence, the primary concern in the case selection is to test whether or not the most distinct domestic institutions will have different E-T policy patterns and outcomes in response to similar E-T development challenges and EU requirements, and thus test whether or not the outcomes are attributed to their institutional arrangements and functions.

The "Varieties of Capitalism" analytical framework is a target-actor-centered approach (Hall and Soskice 2001: 6). Accordingly, the behavior of wind power developers is influenced by regulatory framework promulgated by government agencies, and by the character of organizational arrangements, namely the triangular relationship between target actors, utilities, and their government. This target-actor-centered approach is adopted to construct three comparative grids: regulatory and financial frameworks for wind energy developers, relationships between wind energy developers and other related parties, and wind energy market structures. They are respectively independent and intervening variables, causing the policy outcomes and offering the basis to answers research questions stated above. Comparative studies between these two institutional groups were thus conducted to investigate whether they had different wind energy policy patterns (of arranging their institutions) and outcomes (of energy diffusion).

By using these analytical frameworks, the following questions can be answered. What happened when the liberalized electricity market in the United Kingdom met the burgeoning wind technology industry? Alternatively, has the neoliberal government heavily supported the vulnerable wind energy businesses? Has the prevalent liberalization of the electricity market thwarted German government plans to participate in and prioritize wind energy development? After all, do national governments differ significantly from one another in their policies to promote wind energy development? If so, can those policy differences be attributed to the systematic institutional differences? Are some institutional configurations more effective than others in promoting wind energy? Or do British and German institutional structures each have their own strengths and weaknesses?

The wind energy policies under examination occurred in two stages. The first stage took place in the early and mid-1990s, a period that saw new institutional arrangements for promoting wind power production in response to the need for alternative and clean energy sources. The second stage from the late 1990s to the early 2000s was a period of further institutional improvement aiming at a fully functioning wind development system. Attention will be paid especially to

Table 4.1 Wind Power Developer Relationships with Related Parties

Relations between	The United Kingdom	Germany
Wind Power Developers	Competitive	Cooperative
Developers-Financiers	Horizontal	Vertical
Developers-Employees	Mobile	Integrated
Developers-Utilities	Bidding	Contracted
Developers-States	Arm's-Length	Close

Table 4.2 Comparison of Wind Power Market Structures

	The United Kingdom	Germany
Market Competition	Project Tendering	Project Allocation
No. of Participants	Limited	Large
Type of Product	Less Homogeneity	More Homogeneity
Market Entry and Exit	Free	Restricted
Market Information	Concealing	Sharing

this period because since the late 1990s wind power has been the quickest-developing new energy source globally, achieving an average growth of 30 percent a year.[8]

The evaluations in the next sections will be made according to the institutional differences between the two countries. Such differences are summarized in the tables 4.1 and 4.2. Table 4.1 exhibits wind power developer relationships with related parties, illustrating different business environments for wind power developers in the two countries. Table 4.2 displays wind power market structures, demonstrating the free-market environment in the United Kingdom and the coordinated market environment in Germany.

4.4   WIND POWER POLICY IN THE UNITED KINGDOM

Although the United Kingdom is a net oil and gas exporter, its domestic oil and gas output is expected to decline and its reserves to be used up over the next few decades (even earlier). In addition, the United Kingdom is a coal importer.[9] Owing to the foreseeable overall energy shortage, special regular meetings, known as "Task Force" and its successor "Pilot," between the British government and its primary energy industries have been conducted to search for solutions to the future energy crisis.[10] Wind, as one of the most abundant new energy resources in the United Kingdom, inevitably comes to mind immediately when one thinks of the British government trying to meet its commitments to resolve energy security problems (while gas to replace coal was seemingly the British way to reduce its greenhouse gas emissions over the last decade).

During the 1980s Thatcher era, the neoliberal view of the administration advocated privatization and contracting-out of state assets, and it believed that regulation was a damaging force. In addition, the closure of inefficient coal mines, and the increasing availability of natural gas had accelerated the government's determination to liberalize the gas and electricity market. The 1983 Electricity Act sparked off the formation of an Independent Power Producer (IPP) system. Deregulation of the U.K. natural gas industry began with the Natural Gas Act of 1986. Further to this, the 1989 Electricity Act has allowed nondiscriminatory access to the national grid. These reforms also included the introduction of a pool market in 1990. Meanwhile, over the same period the British state-owned electricity industry was divided into generation, transmission, distribution, and supply, with privatization (especially of the generation part) gradually occurring, and industries being given power supplier choices. In fact, over this period of reform, the British government tried hard to reduce its expenditure and its intervention in the marketplace and thus it was able to slash its energy R&D investments by more than 90 percent. Competition and free market were the government's core strategies to reduce consumer prices and to improve efficiency.

The government changed in 1990 but the energy policy put forward by incoming Prime Minister John Major inherited the dominant neoliberal policy of his predecessor. The electric utility industry was deregulated step by step in the 1990s, working toward full liberalization of the U.K. electricity market in 1998 in line with the 1990 directives of the EU. That is, the 12 English regional electricity boards were to be privatized in 1990, the Scottish generation in 1991, the Northern Ireland board in 1993, other generators in 1995, and nuclear generators in 1996. When the 1996 EU directive for the internal market for electricity was installed, all EU member states were required to open at least 25.37 percent of electricity markets to competition as of February 1999 and increase it to 30 percent as of February 2000.

The Uncoordinated Tendering System

British wind power was developed under such a full liberalization framework. Earlier, there were two departments working for the development of areas unrelated to wind energy. They were the Department of the Environment, Transport and the Regions (DETR) for transport policy, climate change policy, and energy efficiency and CHP policy, and the Department of Energy (DOE) for energy policy.[11] Despite their incumbency for energy (and its efficiency), they were not in a position to take a direct responsibility for sustainable or renewable energy policy. Thus, the United Kingdom had neither strategic schemes for purchasing wind energy nor private companies trading wind power, even though privatization and deregulation of the electricity industry had allowed more room for new energy businesses. Until 1990 only a handful of firms that used electricity themselves could afford the expense of wind powered electricity (Krohn 1998). Even more sparingly, the DOE was disassembled in 1992 and its energy policymaking duties were overtaken by the Department of Trade and Industry (DTI).

Not long before the formation of the DTI, when the electricity supply industry was privatized through the Electricity Act 1989, the Non Fossil Fuel Obligation (NFFO) was devised under section 32 of the act to assist the uncompetitive nuclear industry by imposing the Fossil Fuel Levy (FFL) of 10 percent on the total electricity revenue, from which a small amount was used for renewable energy. Thus, the national government via the DTI formulated the 1990 NFFO for England, the Northern Ireland Non-Fossil Fuel Obligation (NI-NFFO) for Northern Ireland and Wales, and the 1995 Scottish Renewables Obligation (SRO) for Scotland, these were designed to stimulate renewable energy market growth. Under these NFFO/SRO schemes the 12 regional utilities were required to buy a certain volume of renewable energy.[12] Over the initial period of the new scheme for wind energy development, on the government side, the NFFO was administered by the DTI; regional wind resource information was provided by the Energy Technology Support Unit (ETSU), and a fairer market was supposed to be regulated by the Office of Electricity Regulation (OFFER).[13] On the industrial side, the utilities formed the Non-Fossil Purchasing Agency (NFPA) to fulfill their obligations and invited individual power generators to submit tenders for contracts.[14] Along with the liberalized electric power industry, NFFO/SRO was a competitive tendering system adopted to select IPPs to build wind farms and supply electricity to the grid. Basically NFFO/SRO was a market-oriented program. Although it guaranteed a selling price (linked to inflation) and a set number of business years for renewable energy projects, IPPs would be awarded contracts on the basis of the lowest bidding price per kWh. This did not consider site and location issues, which were left to be handled through the Land Use Planning System. In the 1990s, there were five rounds of the NFFO, three rounds of SRO, and two rounds of NI-NFFO. Contracts offered under the first two rounds (NFFO-1 and NFFO-2) lasted for only a few years and terminated on December 31, 1998. Contracts offered under the third, fourth, and fifth rounds (NFFO-3, NFFO-4, and NFFO-5) have been set for up to 15 years. NFFO-3, NFFO-4, and NFFO-5 schemes will continue to be supported until those contracts end. The apparent advantage of this tendering system was the effect of selling price reduction. The wind power price was greatly reduced from some 9p/kWh (NFFO-1) in 1990 to only 2p/kWh (SRO3) in 1998 (chart 4.7).[15] IPPs offered low tender prices under an assumption of declining technology costs in the future. The 12 utilities were to purchase wind power from IPPs who entered such an NFFO/SRO contract and were compensated the difference between the monthly NFFO/SRO price and the power pool market price via the FFL surcharge to consumers (since 2001 NFFO wind power has been auctioned on a half-yearly basis to power suppliers because of the change in law, and therefore the price differences have become narrower and sometimes even profitable).[16]

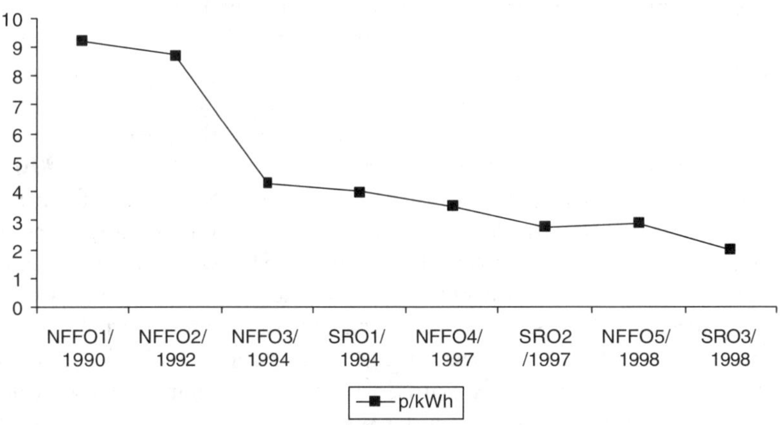

Chart 4.7 Selling Prices of British Wind Power in the 1990s.

Source: Renewable Energy World September/October 2001.

However, these schemes with a minimum government involvement produced many unwanted effects.

First, due to the lack of government directives as to technical standards and, additionally, as to local issues, much time was wasted in project planning and implementation. After winning tender for a contract, the IPPs alone would run into problems in acquiring permits to start the construction of wind farms. It often took up to five years from tendering for wind power projects to commissioning for operation, resulting in many unexpected costs in the process. Each IPP had to manage its own wind survey. For reasons of intensive competition, IPPs were inclined to select roughly the same windy areas. Efforts overlapped in these windy areas and virtually no IPPs were interested in less windy areas that, notwithstanding their not being too favored, were still suitable for wind energy projects. When different IPPs won contracts on locations not far from one another, only one of these producers could proceed because local authorities demanded a minimum "invisible" distance between farms (such that one should not be able to see one wind park from another).

Second, the competitive tendering system limited local participation and created oppositions. The local communities were not competitive enough to bid for wind farm projects that were to be located in their living territories, this lead to conflicts of interest with the nonlocal IPPs. Local and action group oppositions such as the Country Guardian and Cefn Croes at the planning inquiries were common and had seriously interfered with the progress of wind project implementation.[17]

The unexpectedly long planning period had prolonged the break-even period and reduced the projects' profitability for the IPPs who won their contracts for merely 15 years but at a very competitive "bid price." Hence, even though a capacity of more than 2,659 MW of bids had been awarded under the NFFO/SRO scheme in the 1990s, only 391 MW was completed for operation due to the lack of central coordination of the environmental measurement and contract execution (table 4.3). In addition, the NFFO-4/SRO-2 could not boost the wind energy market upon terminating the government investment grant in 1997 (chart 4.8). As far as the degree of state control is concerned, the British model of wind energy policies over this period adopted a combination of competition and laissez-faire. But, as soon as a wind power production tender was accepted, government would "leave the bidder alone" without a significant state involvement, resulting in the poor implementation of the projects, right from singing a contract to final wind energy production.

Table 4.3 Comparison between Contracted and Operating Projects

Order	Effective Start Date	No. of Projects Contracted	No. of Projects Operating	Contracted Capacity Mw Rated (Approx.)	Operating Capacity Mw Rated	No. of Turbines Installed
NFFO-1	1990	9	7	28.0	27.63	75
NFFO-2	1992	49	25	196.0	122.41	355
NFFO-3 (>3.7 MW)	1995	31	9	339.0	88.64	153
NFFO-3 (<3.7 MW)	1995	24	9	46.0	18.14	34
NFFO-4 (>1.76 MW)	1997	48	1	768.0	6.50	5
NFFO-4 (<1.76 MW)	1997	17	4	24.2	6.60	6
NFFO-5 (>2.3 MW)	1998	33	–	768.0	–	–
NFFO-5 (<2.3 MW)	1998	36	2	67.0	3.30	5
SRO-1	1995	11	1	148.0	62.60	110
SRO-2	1997	7	–	101.0	–	–
SRO-3 (>2.3 MW)	1999	11	1	148.0	17.16	26
SRO-3 (<2.3 MW)	1999	17	1	33.0	2.00	1
NI-NFFO-1	1994	6	6	29.0	30.00	60
NI-NFFO-2	1996	2	2	6.0	6.28	9
TOTAL		302	75	2659.0	391.26	839

Source: BWEA Web site via IEA.

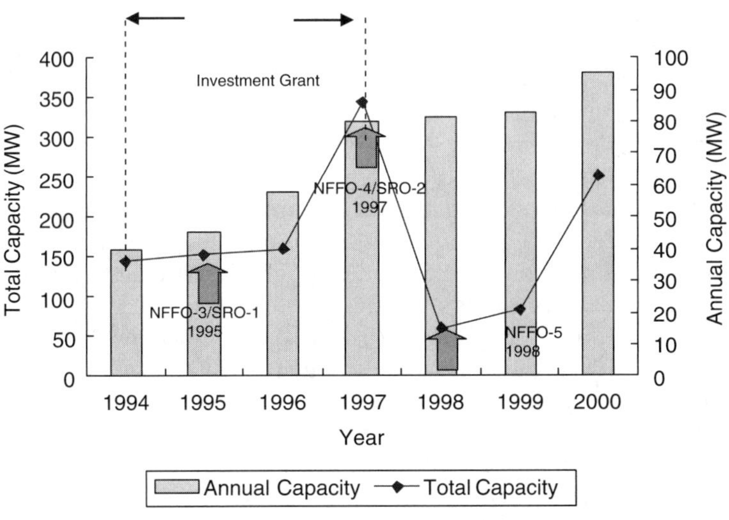

Chart 4.8 British Wind Power Policy Pattern and Outcome in the 1990s.
Source: Wind Energy Annual Report, IEA.

State Involvement in the Early 2000s

To eliminate the unwanted effects of NFFO/SRO, some new rules and regulations were designed. These new arrangements indicated that the British government had exercised four basic state capacities, rather than only market forces, for its wind power development.

First, the government used a "centralized coordination" of wind power investment through a new tendering system: the Merchant Wind Power Initiative 2001 that allowed IPPs to build, own, and operate wind turbines to supply electricity directly to customers rather than via the national grid.[18] The centralized coordination function was not only applied to encourage investment but also to avoid overlapping efforts that greatly harmed the implementation efficiency of the IPPs.

Second, the government effort had also sought more local involvement and ownership of environmental policy since 1999, inviting local authorities to participate in regional strategies and targets for renewable energy. For example, Renewable Energy Systems (RES), a nonlocal IPP that has developed 14 projects in the United Kingdom,[19] in March 2003, announced the installation of two 1.16 MW wind turbines on the northern tip of Scotland near the town of Forss in Caithness in collaboration with a group of local companies, including The Leet Rodgers Practice (Caithness-based architects), NDR Environmental Services (Caithness-based environmental consultancy), and Fivestone Ltd. (an associated company of Abbey Group). Allan Johnston, development executive for RES in Scotland, commented that "the involvement of local companies in the development of the project as a whole has been welcomed and we expect it to be a good example of a successful, small-scale project: developed and built in collaboration with the local community." In addition, dialogues between the government and communities living in wind farm projects areas have been conducted throughout the United Kingdom. For example, the previous Ark Hill wind farm proposal submitted in 2000 was locally publicized by exhibitions, newsletters, press, and meetings with local communities such as affected residents of Strathmore Estate and the Glamis Community Council. Nevertheless, despite these government efforts to involve local communities, there are still some die-hard opposition groups such as local planning authorities (LPAs) that usually complained that the government wanted more wind energy to meet its regional and international commitments but without (or with insufficient) public and local enquires. The national government bypasses them by promoting wind projects over 50 megawatts (MW) in installed capacity, a capacity that according to the requirement of Section 36 of the Electricity Act 1989 needs the DTI's approval instead of the LPA's. For example, the 50 MW Crystal Rig wind farm at Borders, the 60 MW Scroby Sands wind farm at Norfolk, and the 58.5 MW Cefn Croes wind farm at Ceredigion have all been confirmed for commissioning since 2004.

Third, the government has recognized the "unfairness" of the NFFO/SRO subsidization, even though a fossil-fuel levy was imposed to correct the externalities. A new levy, the 2001 Climate Change Levy (CCL), administered by DETR on business users, and from which renewable energy producers are exempted, was introduced. The CCL is applied not only to polluting coal supplies but also to much less polluting natural gas and petroleum gas businesses. This indicates that the government has coordinated the market by financially encouraging through the CCL more development/consumption on nationally targeted energy sources such as wind power rather than gas, even though the latter is cleaner than fossil fuels and cheaper than renewable energies.

Fourth, the government restructured the wind power market system by introducing the 2002 Renewables Obligation cum Green Certificate trading system that improved the old tendering systems of the 1990s. Utilities are now required to fulfill government obligations and regulations, though tradable green certificates were designed to release the obligation burden on them. This compromises the market by indirectly regulating the adoption of renewable energies. Meanwhile, the national wind energy target increased every year from 3 percent in 2002/03 to 4.3 percent in 2003/04, and so on. The four state capacities above are summarized and are compared with the annual growth of wind energy in the late 1990s and the early 2000s (chart 4.9).

On the whole, without central coordination the NFFO and the SRO were unable to perform properly: table 4.3 shows that only 75 out of the 302 contracts became operational and chart 4.8 reveals that the uncoordinated tendering system resulted in an irregular and very unsatisfactory policy outcome. As a result, the British IPPs preferred to spend their efforts more on the well-established gas energy market rather than the primitive wind power industry, leading to a surge of combined-cycle gas energy production in the 1990s. For example, according to DTI, the British demand for natural gas significantly increased by 37 percent between 1993 and 2002. However, in early 2000s, as soon as NFFO/SRO phased out and state efforts were resumed to remove the barriers to wind projects, wind farm installations

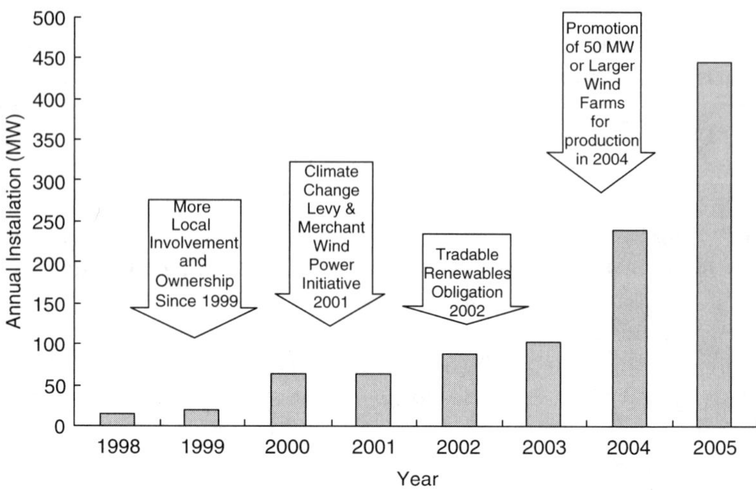

Chart 4.9 Growth of British Wind Energy since Government Involvement in the early 2000s.

Source: BWEA.

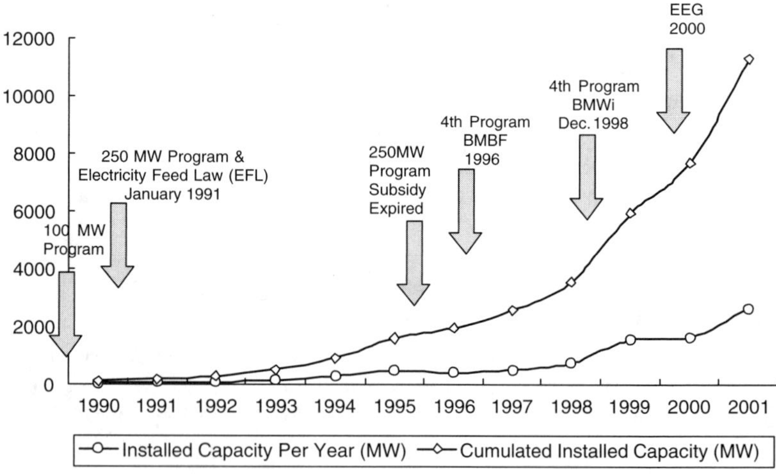

Chart 4.10 German Wind Power Policy Pattern and Outcome.

Source: German Wind Energy Institute.

achieved a sharp increase of 87 MW in 2002 and 103 MW in 2003, 240 MW in 2004 and 446 MW in 2005 (chart 4.9). With government getting involved in the renewable energy market in the early 2000s, the maximum capacity that could get wind power project planning permission greatly reduced the total capacity, for example, from over 295 MW since 2000 to only 24 MW in 2005.

4.5   WIND POWER POLICY IN GERMANY

In Germany, in comparison with the United Kingdom, the national aspiration for new indigenous energy resources aimed at alleviating dependence on single and imported energy sources was held to be obvious and essential. Although Germany is the world's largest and most competitive "brown coal" exporter with one-fifth of global production, it has to import oil and gas. The tight and incomplete range of domestic energy supply caused Germany to look for new energy resources. In addition, its national antipathy to nuclear energy and its support for strong green political parties (such as the Greens) has strengthened the determination of the government to develop renewable energy. However, in stark contrast to the United Kingdom or, the renewable energy policy of Germany over the same decade (1990s) employed its particular industrial policies. The traditional German models of institutions, such as strategic investments, nonprice competition, search for market share, and collective bargaining have been used with good effect by German corporations and policymakers (Hall and Soskice 2001: 21–27). The German government believed that the traditional corporate governance and interfirm collaboration helped to develop new energy resources with a view to quicken renewable energy development and to effectively relieve by the end of the 1990s the forthcoming electricity liberalization price pressure. But, unfortunately, these strategic government efforts in the first half of the 1990s were seriously jeopardized by the strong German electric power industry. Hereunder is the story.

The Response of the Formidable Utilities to Industry-Led "Self-Regulatory" Policies

With the environmental lessons learnt from the nuclear accident of Ukraine's Chernobyl in 1986 and with the popularity of Greens political party in Germany, German people are very enthusiastic about renewable energy. In 1990, a "100 MW-Wind Program" was set without any strategic rules or state supervision. The government delegated the administration and monitoring jobs to industry experts. This was part of the so-called indirect-specific programs for the promotion of specific kinds of innovations, formulated by the then Federal Ministry for Research and Technology (BMFT, *Bundesministerium für Forschung und Technologie*, now BMBF, *Bundesministerium für Bildung und Forschung*) that, in parallel, organized a similar program for five more new technologies, namely, microelectronics, manufacturing technology, information technology, materials technology, and biotechnology, and entrusted industry associations and other non-public organizations with the governance of these programs.[20] However, while the programs for most of these new technologies were proved to be "a very successful means for achieving specific goals within a short period of time" (Vitols 1997 from Ziegler 1989), the wind energy programs were very reluctantly received by the German utilities. The relationship between the government and the utilities became even worse in the following years.

Contemporaneously in 1991, a "250 MW-Wind Program" was designed with the support of an Electricity Feed Law (EFL) to require the utilities to pay a premium (0.06 DEM/kWh) for the wind power to the grids. The EFL took effect from January 1991 and guaranteed fixed, long-term, and high tariffs for wind power producers (usually small and individual ones) selling electricity into the grid, and it required the utilities to absorb 10 percent of wind energy cost (90 percent of the price to be paid by consumers). This additional cost was strongly opposed by those formidable utilities that monopolized the German industry. In 1994 the utilities led by *Preussen Elektra* lodged a complaint against the EFL with both the federal court and the European Commission, claiming that the federal government violated their chartered electricity monopolies bestowed by the Third Reich constitution.[21] Despite the court bringing in a verdict against the utilities, the federal government compromised by amending the EFL in April 1998 asking each level of the utilities to equally share the additional cost by limiting the financial charges of the different utilities to 5 percent each. However, *Preussen Elektra* again confronted the law in the Kiel regional court, requesting the refund of DM 500,000 already paid, arguing that the payment for renewable energy was illegal because the EFL contravened the state aid rules of Article 93 of EC Treaty of the EU. These are typical examples of the overwhelming strength of the traditional institutional structure in the German electric power industry and of how the industry took advantage of the new European laws, and it shows that the federal government did not have the capacity to motivate its utilities. Thus, the initial government efforts aimed at "self-regulating" utilities (without compensation, for example, by the FFL for the British utilities) to absorb the wind power premiums failed. In addition to the refusal to purchase expensive wind power, the utilities were also very reluctant to install wind turbines. Germany is characterized by its cooperative relationship between government and industry (Vitols 1997; Yamamura 1998; MacDougall & Kabashima 1999). But the traditional government-industry cooperation was inapplicable to early wind power projects because the initial small size and local energy were not suitable for large and centralized project developers. The reluctant behavior of the German utilities shows that government-industry cooperation cannot be had when the latter has sufficient power to decline the government's request to comply with undesirable wind energy requirements. Thus, the progress of early wind development was very slow (in comparison with the obviously much faster wind development when a more cooperative government-industry relationship was evidenced).

From "Demanding" to "Obliging" Government Efforts and Regulations

Nonetheless, German state power has not dwindled and efforts have been made to create obedient actors. Local and rural communities were the government's new targets for wind energy development. As a result, more than 90 percent of German wind farms have been owned and operated by private individuals, farmers, or cooperatives since 1989. As the "demanding" regulatory state power requiring utilities to absorb expensive wind energy from small producers without government compensation was facing strong resistance and as the electricity industry guarded their own benefits against the domestic laws by making use of European rules, the German parliament called for a "truce" and decided not to revise the present legislation for a couple of years. Instead, more attention was paid to "obliging" government efforts, including different aspects of market and systemic fixing, hoping that the utilities would accept wind energy more willingly when the costs of wind energy production came down substantially.

First, early wind technology was immature with unsteady wind speed and problematical grid loading of its component and control system. Hence, since 1996, the German government has invested more in R&D funding provided by the "Fourth Program for Energy Research and Technology" (Fourth ERT) implemented by the BMBF. This was not a pure R&D program because the German government worked intensively to arrange a mega database for technical improvement of wind energy. The 100/250 MW programs were coupled with a requirement to report operational data and technical reliability. Many development problems were identified by the government through this arrangement. For example, according to the *Wind Energy Annual Report 2000* of the International Energy Agency (IEA),[22] the grid reinforcement and grid extension costs posed a problem for the German wind energy developers; some 50 percent of the causes of failure were attributed to component and control system failure and around 25 percent of the causes were due to local weather and grid failure. In addition, *Germanischer Lloyd* has provided type approval and standards for wind turbines. These technical rules have not only provided guidelines for technology development but have also boosted protection and confidence of small, individual wind energy developers.

Second, the nontechnical problems were as serious as the technical problems. Since Germany is a very densely populated country, many wind farms either did not have good wind sites or were the cause of complaints by local communities on the grounds of noise, shadow effects, and landscape destruction. To solve these problems, the German government employed its state power, rather than private settlement/argument between wind energy producers and communities in the United Kingdom over the same period, to reset some rules. For example, a rule of minimum 500 m distance from the nearest resident was applied to address civil complaints, privileged status in the building codes was given to the wind energy developers, and performance and operational experience was investigated in the 1998 annual report for the Scientific Measurement and Evaluation Program (WMEP, *Jahresauswertung* 1998, *Wissenschaftliches Meß und Evaluierungsprogramm zum Breitentest* 250 MW Wind) with the contractors.[23] In addition, soft loans through the *Deutsche Ausgleichsbank* and other special tax and accounting arrangements was made available to wind energy developers, but direct funding was avoided.

On the whole, under these technical, social, and financial "obliging" regulations and arrangements, the government has given increased confidence to the market players by establishing a benchmark of profit-making. Particularly, a Wind Energy Conversion System (WECS) under which the market players can calculate whether their new wind energy production business plans can be profitable or not under German meteorological and financial conditions. Accordingly it was generally believed that in 1998 the income generated through selling the electricity from a wind power project would break even at a nominal power for 2,200 annual operating hours at an average wind velocity of between 5.5 and 6 m/s. Therefore, according to the 1997 report of IEA, most of the wind energy projects guided by WECS on the northern windy areas show greater production costs than incomes only for a few years. Under the direction of the government, the growth of wind farms in Germany has been steady and smooth since the second half of the 1990s. According to the records of the German Wind Energy Association (GWEA or BWE, *Bundesverband Windenergie eV*),[24] Germany has since 1997 overtaken the United States to become the world leader in wind power production and has acquired some 18 percent of the global wind turbine market. By December 1997, the cumulated capacity of German wind turbines was 2,081 MW, over six times the U.K. capacity of 319 MW for the same period.

Challenge from the Deadline of the EU Regulations

However, in April 1998, the opening of the general energy market in Germany under the EU regulations was challenging wind energy developers. To satisfy the requirements of EU liberalization of the energy market, the 1998 German retail electricity deregulation law allowed consumers to select their own power suppliers, terminating monopolies of its utilities and permitting wind energy developers access to the utilities' transmission and distribution networks. Thus, the general electricity price went down quickly. The wind power price was jeopardized by falling electricity retail price under such liberalization because these two prices were linked together according to previous laws. To maintain the growth of wind energy development, in December 1998, the newly elected coalition government (the Social Democrats and the Greens) led by *Schröder* salvaged the wind farms by forming a new Ministry of Economics and Technology (BMWi, *Bundesminister für Wirtschaft und Technologie*, now renamed as BMWA, *Bundesministerium für Wirtschaft und Arbeit*) to take care of the fourth ERT.[25] Therefore, in April 2000, EFL was changed to *Erneuerbare-Energien-Gesetz* (EEG) to further strengthen wind energy developers by giving them higher priority of grid connection, guaranteed wind selling price (e.g., 0.178 DEM/kWh for the first five years), and promising them equal absorption of the higher wind energy price across all grid levels. The previous regulation did not consider the different wind resources of various sites. The new rates have however assessed each site's profitability. From then on, windy sites would be compensated for a lower rate of 13.5 pfennig per kWh for a 20-year period, average windy sites for 16.4 pfennig per kWh, and inland sites for 17.3 pfennig per kWh during the 20 years of operation. These new regulations were devised to promote an even development of all possible sites, unlike the United Kingdom where only the windiest sites would be developed. As a result, the German wind energy market has grown steadily under unceasing state efforts (chart 4.10). Unlike the British "deregulation" to reduce electricity price through competition via a tendering system, the German new federal administration heavily coordinated the wind energy market by financially and technically sponsoring rural cooperatives and individuals without a transparent, entrepreneurial IPP system.

Meanwhile, liberalization also sparked mergers and acquisitions in the German electricity industry. For example, Veba and Viag, the nation's third-largest utility, merged into EON; and RWE, Germany's biggest power company, acquired VEW. These amalgamations made the German utilities more powerful than ever. This increased the degree of difficulty in persuading the utilities to purchase more expensive wind energy. Having learnt the lesson in the early 1990s that the monopolized, powerful utilities were reluctant to concede to political demands, the German way of electricity deregulation did not force its utilities to separate their generation, transmission, and distribution so as to maintain a nontensional government-industry relationship. As a result, the six largest utilities remained to act as both transmission grid operators and producers of 85 percent of the country's electricity. The government resorted to negotiation with the utilities to absorb the premium of renewable energy and the latter in turn charged residential consumers for recovering the additional costs they incurred. A new round of wind power development R&D and funding was started and "obliging regulation" put in placed, rather than market-oriented "deregulation" or industry-led "self-regulation," to help the newly emerged, cooperative wind energy developers and to motivate the strongly organized, unwilling utilities. Government coordination in such a subtle form continued in the early 2000s.

4.6   CONCLUSIONS

The aim of this chapter was to examine the delicate and causal relationships between institutional structure and wind energy growth, using Hall and Soskice's "Varieties of Capitalism" institutional approach to compare (1) the wind power regulatory and instrumental frameworks, (2) the wind power developer relationships with related parties, and (3) the wind power market structures, in addition to the different government roles and involvements in terms of market-oriented "deregulation" and industry-led "self-regulation" respectively, in wind energy development projects, between the United Kingdom and Germany. The findings reveal that the U.K. wind energy developers encountered severe developmental problems and local opposition in the absence of state coordination in the 1990s, representing an unsuccessful model to develop wind energy through electricity market "deregulation." However, as soon as state involvement and regulation were resumed to remove these problems and barriers in the early 2000s, the pace of technology diffusion in the United Kingdom became much faster.

In the meantime, the German case indicates that the early industry-led "self-regulatory" form of wind energy policy did not work as effectively as the strongly organized German utilities that made use of their traditional legal right to the exclusive control of the domestic electricity industry and the new European directives to shield themselves from being forced to buy expensive wind energy. In addition, the German utilities became more consolidated by mergers and acquisitions to face the new challenges from European liberalization policies. The German government no longer found its old form of regulation and state power effective and so a new "obliging" form was thus needed. These new government efforts through regulatory change were then found useful to resolve both the technical and non-technical developmental problems. A compromising attitude of the German government toward the ever formidable and defying utilities, a wind power premium sharing arrangement, and governmental guidance for and interaction with small, local, and rural wind energy project cooperatives: all these account for a stable and positive market for wind power developers, finally leading to a rapid progress in wind energy technology advancement and diffusion.

On the whole, the United Kingdom and Germany have both developed wind power at different times under the same European liberalization directives, and they have pursued different courses in accordance with their own institutional structures and state involvements. The regulatory and instrumental frameworks of the two countries are summarized in table 4.4, these frameworks contradict their regulatory regimes and outcomes at different stages. This provides support for the findings of Fioretos (2001) that rudimentary political and economic needs determine a country's options as to how national and European institutions should be integrated. Besides, while the British wind energy developers are entrepreneurial IPPs, the German counterparts are mainly local and rural cooperatives. These different

Table 4.4 Comparative Regulatory and Financial Frameworks

	The United Kingdom	Germany
National Target	Mandatory	Negotiable
Developer Structure	Limited companies	Local cooperatives
Obligation Trading	Renewable Energy Certificate	Nil
Wind Power Price	Floating	Fixed
Obligation Standard	NFFO/SRO 1990s Renewable Obligation 2002	Electricity Feed Law 1991/ EEG 2000
Absorption of Wind Power Premium by Utilities	Compensated by FFL / then exempted from CCL	Shared by utilities (each level up to 5% legally)
Government's Initial Efforts and Outcomes	Market-oriented "deregulation" but local oppositions and contract implementation problems	Industry-led "self-regulation" but utilities refused to adopt expensive wind energy
Government's Later Efforts and Outcomes	1) Coordinating wind power projects to supply electricity directly to customers through the Merchant Wind Power Initiative 2) Inviting local participation in regional strategies and targets for wind projects or bypassing local oppositions via 50 MW projects 3) Coordinating energy market by not exempting gas from the Climate Change Levy 4) Requiring utilities to fulfill regulations through tradable Renewable Obligations	1) Arranging a mega data base and standardizing technical information. 2) Settling disputes between developers and local authorities through clarifying regulations 3) Financial assistance through the *Deutsche Ausgleichsbank* and other special tax and accounting arrangements 4) Giving developers confidence by helping them calculate their profitability through the wind energy conversion system.

British and German corporate governance institutions resulting in different corporate strategies and technology diffusion outcomes reflect the findings of Vitols (2001) who has studied the financial services and pharmaceutical industries of these two countries.

Both the British NFFO/SRO requirements and German governmental control of utilities did not succeed in their E-T developments initially. But the British wind energy market achieved rapid growth in later stages by relieving the pressure on the industry through a new trading scheme. The excellent performance levels of the revised German wind power policies was achieved through supporting and coordinating networks of nonutility companies in the second half of the 1990s. This reflects findings of Wood (2001) that dirigiste policies and implementation strategies that use only sanctions and/or incentives do not work; instead, policies are more effective if the state "heightens" its institutional comparative advantages, as Casper and Kettler (2001) have shown in their biotechnology case studies.

In an era of regionalization and globalization, interactions between national and international institutions are inevitably creating more tensions. Nevertheless, the findings in the chapter lead us to question whether "deregulated" market forces or devolved industrial governance or highly "regulatory" state power is an effective method for individual nation-states to survive conflicts of interest arising from using/manufacturing expensive wind energy and increasing cross-border activities. Instead, the later development stages of both cases show that a combination of market forces and state power was the solution to problems faced in the initial stages of wind energy development. It must be made clear however that the fundamental force behind developments in the United Kingdom was still market-led, and in Germany it was the deployment of state power.

The general lessons we have learnt from the cases are about the dynamic role of governments and the necessity of institutional and regulatory adjustments to accommodate changing conditions. The "varieties of capitalism" analytical framework and "the regulatory state" theoretical perspective, both are useful for understanding each economy's specific capacities but have little application for improving institutional and regulatory comparative advantages. By studying the wind energy industry, the findings reveal how governments guided their industries/societies to a stronger developmental path with special institutional and regulatory arrangements that can reverse the wind energy industry's disadvantaged situation. These special institutional arrangements are institutional anomalies and contradictions that coexist alongside their own dominant institutional patterns. That is to say, "varieties of capitalism" can deliberately be arranged by tweaking institutions with an *elastic* form of regulation beyond the traditional rigid rules of legal liability within a country in order to cope with different challenges in different sectors.

The intervening variables between regulations and wind energy diffusion are all-inclusive state-societal structural changes and "obliging" government efforts that include from greater participation by local communities, technical and nontechnical information dispersion, conflict mediation, market and systemic failure fixing, confidence building, and developmental cost and risk sharing, in combination with the increasing market force of EU liberalization. The transformation from market-oriented deregulation or industry-led self-regulation or demanding regulation to "obliging" regulation and state involvement is demonstrated by the state's efforts to change a "static" (de)regulatory or self-regulatory framework to a "dynamic" one through more institutional interactions between government, industries, and local communities. However, it must be admitted that Germany employed its state-societal structural change and "obliging" regulation and state involvement *much earlier* than the United Kingdom, this explains the much earlier (and quicker) diffusion of wind energy in the former than the latter.

Most environmental technologies such as wind power generators are not yet technologies for promoting higher productivity and efficiency and their markets are generally restricted by the barriers of E-T standards and user-skill distributions. In fact, they are long-term, risky, incremental innovation and capital goods. Thus, they are temporarily not attractive to companies, particularly those financed by volatile capital markets that seek profit maximization and are inclined to engage in technologies that can enable them to immediately increase productivity and efficiency. Industries in some countries, especially those exposed to policymaking and implementation determined by a pluralistic parliamentary or congressional voting system and arm's-length state-industry relationships, tend to lobby or repeal unfavorable laws and regulations and/or invest only in switchable or fast-moving assets and technologies to achieve radical innovation. Further, if the worst comes to the worst, when E-T policymaking, implementation, and rectification are carried out through a powerless bureaucracy, governments will have weaker state capacity to influence this corporate behavior. Therefore, although a considerable degree of "obliging" regulation is employed, their different state-societal structures and relationships will have different approaches and patterns for fixing their own development problems, leading to different policy outcomes. The future of wind energy innovation and development influenced by German/U.K./European policy will be contingent also upon local state-societal structures and relationships.

Studying the development of new technologies in conflict with intransigent social and/or industrial interests is a significant line of research for further exploring the virtue of state efforts within different institutional structures. It would be beneficial to study the developmental trajectory of another domestic policy sector under similar degrees of regulations (most countries have to heavily regulate their actors in this sector) but different state-societal structures: that is, the packaging waste recycling industry, a case study in the following chapter.

5

Varieties of State-Societal Structure: Packaging Waste Recycling Development in the United States, the United Kingdom, Australia, Germany, Sweden, and Switzerland

5.1   Introduction

Given the obvious benefits of waste recycling over landfill disposal and incineration, and the fact that the latest technologies can recycle virtually all types of garbage, the last two decades have seen a good many advanced industrial countries strive for higher packaging waste recycling (PWR) rates by evolving new plans to solve developmental problems.[1] These PWR developmental problems are similar to those of other environmental technologies, ranging from poor productivity, market and systemic failure, conflict of interest, low commercial value, lack of infrastructure, unfair politicization, and insufficient local participation.

In order to address these PWR policy problems, governments are inclined to employ a *higher* degree of state power through laws and regulations that, firstly, may encourage producers to adopt/design more recyclable containers and, secondly, may prohibit consumers from mishandling recyclable waste and motivate them to return their consumed recyclable containers to curbside/designated collection places from where they are sent to recycling processors. Yet, how can governments regulate private corporate and individual affairs to achieve their PWR goals without stifling the industry's competitiveness? In this regard, do PWR laws and regulations or any other factors matter?

Copious studies have examined the effects of different PWR laws and regulations, particularly between (more regulated) container deposit legislation (CDL) and (less regulated) curbside recycling systems (CRS) (e.g., Hatch 1990; Lenehan 1992; Aumônier & Troni 2000; Wilson 2002).[2] Obviously those with a higher degree of state power through laws and regulations and those without it have very different PWR rates overall. Yet, though these studies reveal that most of the recycling rates in CDL states/cities are significantly higher than those in CRS state/cities, they cannot explain the recent decline in recycling rates in some CDL states. For example, many American CDL States have had their PWR rates fall more than 10 percent since the mid-1990s, but the increasing recycling rates in some non-CDL states (e.g., the PWR rates of many one-way beverage bottles in Germany, Sweden, and Switzerland, which suspended the use of CDL) rose sharply to nearly 90 percent in the late 1990s. This implies that, not only do regulations matter, different political economies may also have different impacts on PWR development. Thus, a comparative study of this highly regulated and government-led policy sector (i.e., CDL) across a range of state-societal structures in liberal market economies (LMEs) and coordinated market economies (CMEs) may determine what sort of domestic institutions can generate a positive outcome in terms of boosting PWR rates.

Indeed, what has happened in these PWR "leading" and "laggard" countries? To what extent have the governments established regulatory frameworks to govern the behavior of their packaged goods producers, retailers, and consumers? Have they established rules and regulations to coerce producers/bottlers into adopting recycled materials and recyclable containers? If not, what has constrained them from (further) regulating industries? If so, what was the response of these industries and what were the outcomes? Finally, what determines whether institutions were an enabling or constraining influence on the development of this particular industry?

Energies have been applied to diversified institutional studies. Besides arguments for laws and regulations (e.g., Vogel 1996; Busch 2002; Grossmann 2003), studies from a broader institutional perspective show how norms and organizational arrangements may also have a significant impact on national development (e.g., Hall 1986; Powell and DiMaggio 1991; Pempel 1998). I will hereafter demonstrate how norms and organizational arrangements in terms of state structure and its relation to society produce laws and regulations for PWR development. In this particular sector, laws and regulations have been widely employed and it is therefore of primary importance to study how they have been formed through legislative processes with their own norms and organizational arrangements. It appears that norms and organizational arrangements in the PWR law-making arena are respectively legislative traditions and government structures in which government executives carry out considerable reforms to motivate their industries and consumers to increase recycling rates. Different legislative traditions and government structures may condition different legislative processes and executive powers to formulate and implement PWR laws and regulations.

The purpose of this chapter is to study in detail the extent of their influence. In doing so, the institutional content of legislative traditions and government structures in terms of interaction between governments and societal groups in the context of the PWR law and program making and implementation will be scrutinized, arguing that, not only regulation but state structure and its institutional relation to society, as well, matter to PWR outcomes.

For this study of different institutions, commonalities and variations of state structures in terms of *pluralistic* legislative and executive powers and their relationships (within which PWR laws and programs are designed and made) and government-industry relationships for PWR development between "leading" and "laggard" countries are examined. This comparative study will advance two propositions. First, differences in legislative tradition and government structure, particularly in the relationship between the government executive and legislature on different levels, give rise to diverging national and local enactments of PWR laws and regulations. Second, in the process of PWR law and program-making, the "opposing" or "discordant" legislature and "majoritarian" executive produce constraints that limit the development of the PWR industry, while the "compromising" or "collaborative" legislature and executive governments formed by a "coalition" exert enabling effects that boost the rates of PWR.[3] This will be illustrated by comparing the contexts of six sampled countries, that is, the United States, the United Kingdom, Australia, Germany, Sweden, and Switzerland. But first we look at the comparative figures and methodology.

5.2 COMPARATIVE FIGURES AND METHODOLOGY

Although the case studies in the last chapter have proved the importance and the specific nature of regulations, under the concept of the "different traits design" explained in chapter 3, any policy sector that was characterized by a high degree of regulation and yet but produced inconsistent policy outcomes should be scrutinized. Poor performance of highly regulated PWR industries in some advanced industrial countries has thus fallen into this "chasm" to be examined. The examination will begin with a discussion of comparative figures showing that (highly regulated) CDL states' rates are much higher than those of their non-CDL states in the Anglo-Saxon countries. For example,

- Michigan, California, Massachusetts, Oregon, and New York have employed CDL for more than a decade. Their beverage container recycling rates are almost twice as high as the national rate of about 44 percent (chart 5.1). According to the Container Recycling Institute, the 40 states without CDL are at an average rate of about 30 percent. Similarly, South Australia is the only CDL State in Australia, it has by far the highest rate of container recovery of about 85 to 95 percent compared with the much lower rates in its other non-CDL States/Territories, for example, 45 to 55 percent in New South Wales and 32 to 64 percent in the whole country (table 5.1).

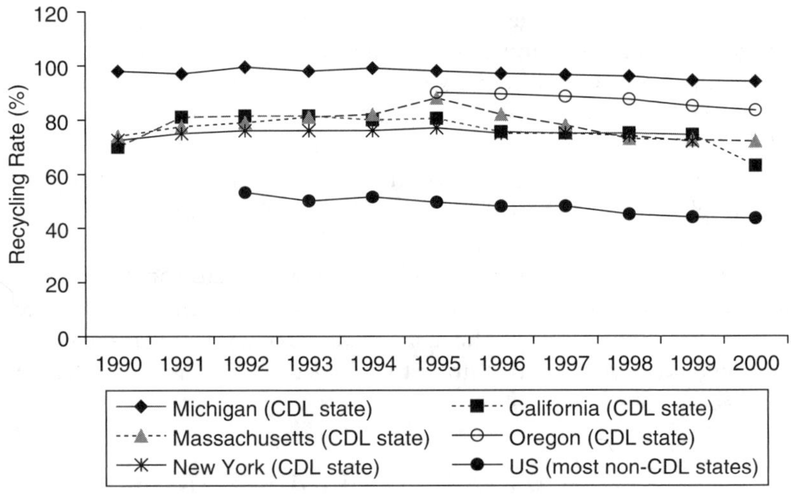

Chart 5.1 National Beverage Container Recycling Rates in Selected CDL States in the United States.

Remarks: Figures include aluminum cans, nonrefillable glass bottles, and PET bottles.
National recycling rates based on tonnage not tonnages.
Source: Container Recycling Institute.

Table 5.1 National Container Recovery Rates (%) in Selected CDL and Non-CDL States of Australia

	SA 1991 with CDL	SA 1997 with CDL	NSW 1996 with Non-CDL	Australia 1990 in Average	Australia 1998 in Average
Glass	84–89	83	55	25	44
Aluminum	85	82	40	31	64
PET	54	70	45	Nil	32*

*Figure in 2000.
Source: EPA SA, EPA NSW, Scoping Paper for Beer and Carbonated Soft Drink IWRP (1996 figures derived from BIEC Data). Plastics and Chemicals Industries Association Plastics Recycling Survey 2001, ACI Glass Packaging 2000, Aluminium Can Group 2000.

Further, some West European countries significantly surpassed Anglo-Saxon countries in PWR rate and had more rapid growth patterns in the last two decades. For example,

- According to the 1996 Report of the *Glass Gazette of the European Glass Federation*, glass recycling rates in Germany, Sweden, and Switzerland were 79 percent, 72 percent, and 89 percent respectively, while those of the United States, the United Kingdom, and Australia were respectively 33 percent, 22 percent, and 44 percent only (chart 5.2); according to the 1998 statistics of the European Aluminium Association, aluminum can recycling rates in Germany, Sweden, and Switzerland were 86 percent, 87 percent and 89 percent respectively, but those of the United States, the United Kingdom, and Australia were respectively 53 percent, 38 percent and 64 percent only (chart 5.3).
- According to the American Aluminum Association, the aluminum can recycling rate fell gradually from 65 percent in 1992, to 55 percent in 1999, and to 48 percent in 2002. According to the American Plastics Council, the PET bottle recycling rate fell gradually from 38 percent in 1994 to 23 percent in 1999. According to the U.S. Bureau of the Census, the glass bottle recycling rate fell from 32 percent in 1994 to 30.5 percent in 2000. In contrast, according to the Swiss Federal Statistical Office, the Swiss aluminum can recycling rate increased from 40 percent in 1990, to 80 percent in 1994, and to 89 percent in 2000. The glass bottle recycling rate increased from 65 percent in 1990, to 80 percent in 1994, and to 89 percent in 2000. Furthermore, according to the *AB Svenska Returpack*, the Swedish aluminum can recycling rate increased from 63 percent

in 1984, to 75 percent in 1987, and to 86 percent in 2002. The PET bottle recycling rate increased from 51 percent in 1994, to 78 percent in 2001. On the whole, the major PWR development in these Anglo-Saxon countries has been regressive or still at lower rates in comparison with the progressive and higher rates of the European countries (chart 5.4).

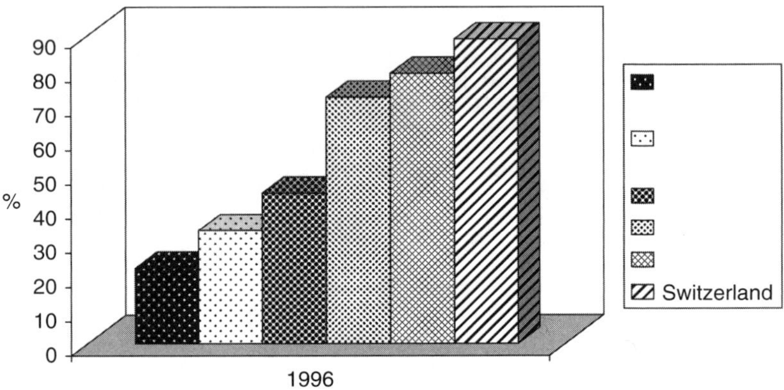

Chart 5.2   Glass Container Recycling Rates across Political Economies.

Source: Glass Gazette, September 1997, the European Glass Container Federation Glass Container Institute.

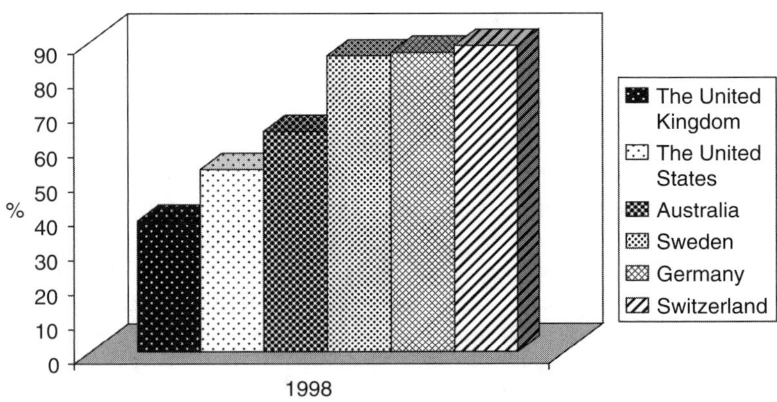

Chart 5.3   Aluminum Can Recycling Rates across Political Economies.

Source: The European Aluminium Association, the Aluminium Can Group, and State of the Environment Australia.

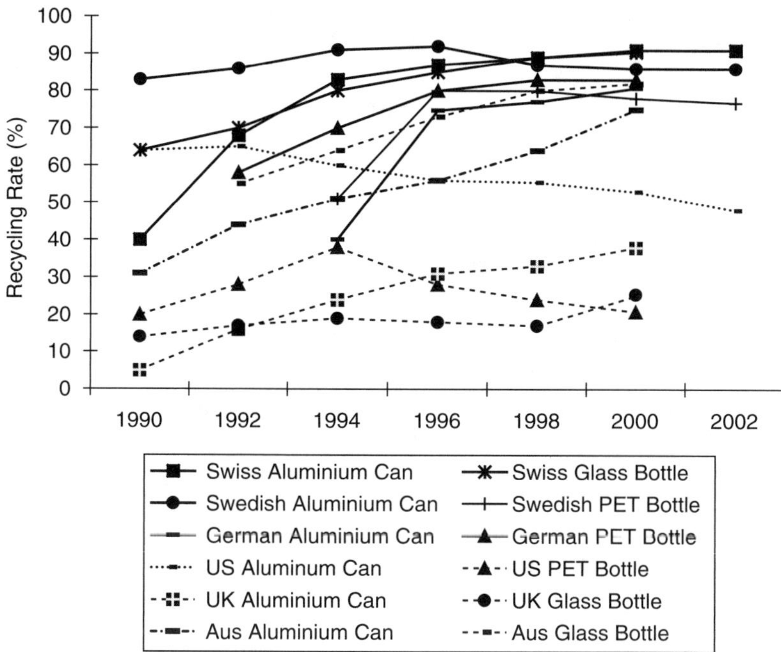

Chart 5.4 Packaging Waste Recycling Rates between Anglo-Saxon LMEs and European CMEs.

Source: Container Recycling Institute; The Aluminum Association Inc., United States; USEPA Office of Solid Waste; American Plastic Council; Glass Packaging Institute, U.S. Census Bureau; BioCycle; Swiss Federal Statistical Office; *AB Svenska Returpack*; *Gesellschaft für Verpackungsmarktforschung mbH*, Germany; *Bundesministerium für Umwelt, Naturschutz und Reaktorsicherheit*, Germany; Beverage Industry Environment Council, Australia; Aluminium Can Group, Australia; IGORA, Switzerland; and Department for Environment, Food and Rural Affairs, United Kingdom.

This view shares the basic argument in the last chapter that (re-) regulation matters. But on close examination, one would notice that the PWR rates of the highly regulated CDL Anglo-Saxon states began to drop significantly soon after operation (see charts 5.1 and 5.4). To explain these contrasting domestic figures and divergent national PWR policy outcomes, institutional structures and bills of CDL between three "laggard" (the United States, the United Kingdom, and Australia) and three "leading" (Germany, Sweden, and Switzerland) beverage container recycling countries will be compared. This sample selection is more or less based on the conundrum that the majority considered the employment of CDLs at the beginning but have had different PWR outcomes. The comparison is also based on the grounds that they are at the frontier of making PWR laws and regulations, having at least a decade of policymaking experience with considerable and noteworthy track records. The beverage container as one of the most significant and popular packaging waste types in societies is brought into focus to facilitate the search for actual state-societal factors in the making of PWR "leaders" and "laggards." One might suggest that the more compact a country is, the more willing consumers are to return their consumed bottles. Yet, this matters little, not because it is very hard to select places with very similar physical particulars, but because data show that size and population density issues are not major factors in the different performance of PWR development.

5.3   THE UNITED STATES

The American executive and legislative (and judicial) bodies have clearly separate powers. The U.S. president is both the chief of state and the head of government, but the president's power is separated from that of Congress as the president and the Cabinet members cannot introduce bills in their names, even though the executive usually initiates draft legislation and has the power to call bills into special session if Congress does not act on them. Rather, congressmen have the right of legislative introduction and have a strong position in making, changing, and repealing laws. These works are usually undertaken through bills to be examined and modified (or rejected) by committees and then voted on by legislators. Thus, the executive cannot make policies within its own discretion without the majority support of the legislature if the president's own political party has not dominated Congress. Alternatively, the executive may direct orders to various federal agencies or departments to help perform their duties, but Congress may also overturn them when it deems it to be in conflict with the law. Although the president can veto such a decision, Congress can overturn his veto by a two-thirds majority of both houses (Laundy 1989: 32). Thus, the ultimately overriding power of the legislature has kept the executive from full control of PWR laws and regulations. For example, in 1978, the Federal Trade Commission ruled the soft drink bottlers' exclusive franchise to be a breach of the antitrust law and requested them to include considerable volume of refillable bottles in their production.[4] However, the bottlers resorted to asking Congress for protection. Finally, in 1980, a Soft Drink Interbrand Competition Act was overwhelmingly passed by Congress that then stood on the side of the business sector to reject the executive's legal requirement to use refillable bottles, leading to a decline in their use from virtually 100 percent in 1958 down to less than 2 percent in 1995. For another example, as of April 2003, in Massachusetts, Governor Romney made a proposal for revamping the "bottle bill" to include a wider selection of beverage containers for the existing CDL in his 2003–2004 budget recommendations, but the proposal was unfortunately rejected by legislators.

Further, U.S. legislators of a minority party can initiate bills without being influenced by the executive and their legislative caucuses. In fact, private members of Congress have quite often proposed their own PWR bills. For example, even though the Republican Party controlled both chambers of Congress and the White House in May 2003, Senator Andrea F. Nuciforo, Jr. (D-Pittsfield), a Democrat, introduced a bill to include approximately 700 million noncarbonated beverage containers presently exempt from the 5-cent deposit-refund scheme in Massachusetts. In fact, in the years from 2001 to 2002, of the 24 bottle bills, 16 were initiated by the Democrats. In addition, even party discipline and solidarity do not exist in the United States (Laundy 1989: 70). A private Senate member without his or her political party's support can also propose a law. For example, Senator Jim Jeffords of Vermont renounced his Republican Party membership (as an independent Senator) in early June 2001 and introduced the National Beverage Producer Responsibility Act of 2002 (and 2003), which would require the beverage industry to collect a 10-cent deposit on every beverage container at a national level.

Due to the power shared with the federal government stipulated by the Constitution, each U.S. State may have its own recycling legislature independent from federal control. The federal government cannot employ a cohesive mandate requiring every State to employ similar PWR laws and regulations. When the federal government put forward the New Federalism using block grants in the 1980s and 1990s, it made the State governments more inclined toward CDL where collection works could be shifted to industries and consumers and thus local governments could reduce their financial responsibility.[5] However, although CDL proposals have emerged in many State governments, the decision of local legislatures is subject to lobbies and struggles between the proponents (usually from environmental activist groups) and opponents (mostly from industries). As a result, only a handful of States have enacted harsh PWR laws and regulations different from the others. By the early 2000s, only 10 out of the 50 States (and one district)—Connecticut, Iowa, Vermont, Maine, Michigan, Delaware, Massachusetts, New York, California, and Hawaii—have adopted CDL, while the others have declined it. Even filed or prepared CDLs have been continually challenged by the industry (Ohio and Washington in 1979, Colorado in 1982, Oregon in 1996, Massachusetts on several occasions, and New York, Iowa, Maine, and Connecticut recently).[6]

Apart from Congress proposing and voting in bills, industry and community groups are also entitled to propose referendums for making, changing, or repealing laws. In fact, PWR legislation by community initiative has been very popular in many U.S. States. The availability of local initiative has enabled both environmental activists and industries to establish or repeal laws against each other. This can be typically illustrated by the process of deposit legislation in Columbia, Missouri. The first American deposit law in this municipality was enacted purely through *civic initiative* and *referendum*. In 1976, a joint-force of civic and environmental activists, including the League of Women Voters, the Audubon Society, Missourians for Safe Energy, and the Sierra Club, wanting to set up a campaign under the name of Columbians Against Throwaways (CAT), initiated the first bottle bill in the country. This initiative tried to make Columbia in the State of Missouri the pioneer municipality in the United States, introducing the Columbia's Beverage CDL in 1977 for reusing and recycling beverage containers. At first, when the Columbia City Council hesitated about the proposed ordinance, CAT acquired sufficient signatures to support their petition and resubmitted it to the City Council. However, the Council was still very reluctant to agree to it and thus CAT turned to the public voters. The ordinance was finally passed by a referendum with a narrow margin of 53 to 47 percent. Yet, the beverage and grocery industries unceasingly opposed the law and tried hard to repeal it. Even though it was a long battle with four failed repeal attempts, it finally conducted a first successful attempt to lift the deposit ordinance through a referendum in early 2002.[7] Following this precedent case, the CDLs in Maine and Michigan in 1976 and Massachusetts in 1982 were also enacted through referendum.

The federal system and the independent legislature in the United States has allowed and encouraged decentralized and non-government-influenced PWR legislation. The Federal agencies were only able to manage the PWR in the country via the broad authority given by Subtitle D of the Resources Conservation and Recovery Act of 1976 (amended from the Solid Waste Disposal Act of 1965) to regulate the disposal of nonhazardous solid wastes and to encourage States to develop comprehensive plans to manage nonhazardous industrial solid waste and municipal solid waste. The federal role confined itself to establishing national goals, offering guidance and developing educational materials. Ironically since November 2003, after disappointedly limited federal government efforts on PWR, a national CDL has been introduced in the U.S. Senate primarily by Senator James Jeffords (I-VT), it was cosponsored by John F. Kerry (D-MA), Patrick J. Leahy (D-VT), Joseph I. Lieberman (D-CT), and Daniel Akaka (D-HI). All these five senators come from states with CDLs. Yet, they have to contest every inch of the ground against the others in the "opposing" or "discordant" American legislature.

Overall, the institutions in the American state structure do not favor the makings of a concerted PWR policy, let alone the implementations of a national PWR policy.

5.4   THE UNITED KINGDOM

The British legislature's pressure is generally less exerted on its executive because, unlike the American state structure, the latter is composed of the prime minister and his/her cabinet ministers who are always the members of Parliament (usually the lower house). They are politically supported by the majority in the popularly elected House of Commons, otherwise they are obliged to resign (Laundy 1989: 30–31). Thus, there is apparently no major separation of executive and legislature in the United Kingdom. For this reason, the British executive seems more powerful than its U.S. counterpart and is thus in a better position to control the PWR legislation. Other than the executive's primary legislation (Acts of Parliament), ministers may put forward devolved or secondary legislation (parallel statutory instruments) to be endorsed by legislators under powers given to them by an Act. For example, the statutory instrument of purchasing packaging waste recovery notes (PRN) either directly themselves or through compliance schemes under the Producer Responsibility Obligations (Packaging waste) Regulations (as an SI 1997 No. 648 and amended in SI 1361 and SI 3447 in 1999) (PRO) was established by executives by exercising the powers conferred upon them by sections 93 and 94(1)(c) of the Environment Act 1995.[8] This statutory instrument is further interpreted by the Office of Fair Trading as regulations to be enforced by highly competitive compliance schemes. As a result, a total of 26 packaging waste compliance schemes, the two largest ones being Valpak and Wastepack, have been fiercely competing in the PRN market with one another in the country. Further, the second legislative power also allows the executive to adjust the increase of recycling and recovery rates on a yearly basis. However, although the British executive seems more powerful than its U.S. counterpart, the legislative approval is still fundamentally influenced and made by the whole legislature that includes not only those from the majority ruling party but also those from the opposition parties. Consequently the British executives are in a better position to control the PWR legislation but are still fully responsible to the Parliament.

Also, unlike the United States, the United Kingdom does not have a federal system of government but is a unitary state with a series of devolved parliaments and assemblies for Scotland, Wales, and Northern Ireland (there were elections for a new Scottish Parliament and a new Welsh Assembly in 1999, and since October 2002 the Northern Ireland Parliament has been suspended).[9] Yet, the devolved parliaments have no constitutional right to exist and can be unilaterally revoked by Acts of the central Parliament. This concentration of the Conservative government's power on promoting a radical program of privatization and deregulation that sharply contrasted with the principle of government involvement and strong CDLs explains the absence of harsh PWR legislation in the United Kingdom over the last 20 years. The regulatory vacuum lasted for decades until the Labour government came into power and the PRO cum PRN system was finally in place. Based on these, a forward market trading system has been introduced since 2000, companies who have recycled more than the required amount may receive PRNs and can sell their PRNs to others who do not meet the target or who export PRNs.

Under British traditions, except for expenditure or taxation, bills can be initiated also by any private parliamentary members or backbenchers (Laundy 1989: 68–69). Therefore, although the policy preparation work is dominated by the executive (generally backed up by most members of the legislature), the policymaking stage has from time to time been influenced by backbenchers in cases where the interests of their constituencies were at stake (Wood 1991: 121). Despite the fact that the proportion of the number of Private Members' Bills passed in the total legislative output in the United Kingdom has always been very low, these have still provided a possible channel for private members to dodge the control of the executive. For example, the Household Waste Recycling Bill introduced by backbencher Joan Ruddock requiring all local authorities to offer curbside collection of at least two recyclable materials by the end of 2010 was surprisingly passed at its second reading in the House of Commons in March 2003, although the government had refrained from backing the bill but changed its mind at the last moment.[10] This "loop-hole" in parliamentary mechanisms may therefore cause part of the PWR laws and regulations in the country to be out of line with the government's policy; and thus British constitutional traditions endow any parliamentary member the right to initial "opposing" or "discordant" bills, though not as often as its American counterpart.

The British government put forward PWR regulations by offering different options for individual brand owners to register through 26 compliance scheme companies (or directly with government agencies but at a slightly higher cost), regulations that were to be implemented by England, but not necessarily by Scotland, Wales, and North Ireland, these were subject to the acceptance of their newly established parliaments/assemblies since the parliamentary reform in the late 1990s. The participants tend to be motivated by competitive terms and services offered by the compliance schemes. These schemes are for privately owned profit-making companies that are expected to compete with one another. They should be seen as open and fair through arm's-length, transparent trading systems.

In addition, in order to meet the requirements of the EC Directive on Packaging and Packaging Waste 94/62/EC, a nondepartmental public body, the Advisory Committee on Packaging (ACP), was created in 1996 to advise the government on regulations implementing recovery and recycling targets. This made the government more informed to put forward realistic requirements and amended the recovery and recycling targets for 1999 and 2000. For example, Schedule 2 (recovery and recycling obligations) to the Producer Responsibility Obligations (Packaging Waste) Regulations 1997 was amended as follows: "in paragraph 4(b), for '43%' substitute '45%'; in paragraph 5(b), for '11%' substitute '13%.' " Also, having experienced a need to have a more intensive dialogue with the industry, the ACP was restructured in 2002 by inviting one representative from each main industrial actor, that is, packaging raw material manufacturers, converters, packer/fillers and sellers. In addition, government officials from the Department for Environment Food and Rural Affairs (DEFRA), the Department Trade and Industry (DTI), the Environment Agency (EA), Scottish Environment Protection Agency (SEPA), the Scottish Executive and the Welsh Assembly are also summoned to discuss with the representatives of the four industries. Thus, a state-societal conversational mechanism was finally established.

5.5 AUSTRALIA

Australia has similar law-making procedures and traditions as compared with those of the United Kingdom. The federal executives, namely, the prime minister, and his cabinet, are from the party with a majority in the House of Representatives and they introduce the great majority of bills in Parliament. In addition to primary law initiation, these executives can also make secondary legislation, called Regulations, by ministerial decree. However, executive power is sometimes counteracted by a more powerful upper house (elected members with considerable legislative power) than the British model (unelected members and limited legislative power). Nongovernment members in the lower house are also very active. There has been a significant rise to the number of Private Members' Bills recently. Between 1901 and 1988, 104 Private Members' bills were introduced into the House, but by the end of 1995 this number had increased to 175.[11] In addition, like the United States, Australia has a federal system of government consisting of a central Parliament (Commonwealth or Federal Parliament) and a number of regional parliaments, which are responsible for their own regional and prescribed legislatures under the Australian Constitution. Such a combination of the U.S.-style federal system characterized by its powerful upper house and the Westminster-style parliamentary system may cause more disagreement between the executive and the upper house, and conflict of power between the federal and the state/territory governments. In fact, the power sharing between the federal and local governments and strong elective upper house has considerably hindered the national executive's powers (Weller and Fleming 2003: 16). As a result, the Australian States, like the American States, are free to choose their own PWR legislations, leading to the enactment of the highly regulated CDL in South Australia (Beverage Container Act 1975 and the Environment Protection Act 1993) different from the CRS or other less-regulated PWR legislations in other five States and several Territories. These facts have demonstrated the partial existence of the country's "opposing" or "discordant" legislature at different levels, though among the three Anglo-Saxon cases it is the least here in Australia.

Yet, the federal executive has tried very hard to increase its powers in recent years by imposing national regulation through international agreements and conventions to increase the federal government's power and give it authority to act over the states. (Moon and Sharman 2003: 18–19). Since 1991, the federal government has become involved with waste management policies for reducing packaging waste in its States and Territories through the international Australian and New Zealand Environment and Conservation Council (ANZECC) by developing the 1991 National Packaging Guidelines, the 1992 National Waste Minimization and Recycling Strategy and the 1992 National Kerbside Recycling Strategy with National Material Specific Waste Reduction

Agreements. Driven by these cross-border momentums and the new membership of Papua New Guinea in 1998, the Commonwealth struck a deal with the Australian States, the Australian Capital Territory, the Northern Territory and the Australian Local Government Association known as the Intergovernmental Agreement on making joint legislative provision for establishing a body to determine national environment protection measures. Supported by this National Environment Protection Council Act 1994 and the corresponding Acts in each State/Territory, the ANZECC further regulated the national PWR issues by formulating the National Packaging Covenant (NPC) in 1999, of which the NPC is a self-regulatory agreement between industries and governments, based on the principles of shared responsibility through product stewardship to manage the disposal of waste packaging. Brand owners who are not prepared to sign the NPC have to alternatively fulfill their obligations under the National Environment Protection Measure on Used Packaging Materials (NEPM) by undertaking, demonstrating, and reporting packaging waste recovery and reutilization specified by agreement among jurisdictions. The NEPM has become a policy in all State/Territory governments. To facilitate this, a NPC council is formed, consisting of government officials at federal and local levels and industrial representatives. The senior executives of Federal, State, and local government thus meet regularly with industrial representatives, for example, the Beverage Industry Environment Council (BIEC) and the Australia Council of Recyclers (ACOR), to discuss issues in packaging and recycling industries that have to regularly submit PWR progress reports to the NPC Council.[12] Although the market-oriented national policies of the NPC does not regulate companies' packaging type and waste collecting way, this cooperation between the federal government and industries over PWR issues has seemingly helped make federal PWR policies and considerably overturned the independent PWR law and regulation-making power of the State governments.

5.6 GERMANY

The German state structure has contributed to a strong position of the executive, chancellor, and his cabinet, in the lower house (*Bundestag*). Usually these executives form a coalition government in which the chancellor comes from the party with the most seats in the *Bundestag*, while the vice chancellor comes from the smaller party in the coalition. They are empowered to formulate policy guidelines and initiate most legislative proposals. Due to the fact that 50 percent of *Bundestag* members must be politically elected (the other half are geographically elected) and that a political party must have a minimum 5 percent of the vote or three directly elected deputies to be represented in the *Bundestag*, executives always enjoy a high degree of political loyalty and steadily control caucuses in the *Bundestag* without political fragmentation, and minor parties cannot be strong without representing a substantial constituent. The substantial legislative work is done by committees between three reading debates. Although the opposition parties may scrutinize bills during question hours (*Fragestunde*, organized by the *Bundestag*), so that government representatives can respond, the time in the process provided for constituents is usually deemed insufficient. Ministry officials are responsible for elaborating new bills and preparing memorandums for the minister to address it in the legislature. Then the bill is officially presented to the Parliament. Since the 1977 German Waste Law, the majority of PWR bills proposed and/or passed in each *Bundestag* were government bills.

Although both the *Bundestag* and the *Bundesrat* (the upper house) may propose legislation, only 6.6 percent of all bills introduced and 3.3 percent of all Acts passed at the federal level between 1949 and 1994 originated in the *Bundesrat* (Saalfeld in Norton ed. 1998: 46–48), unlike the active bill initiation in the American upper house. Traditionally, the power of the *Bundesrat* is subordinate to that of the *Bundestag*. In 1949 only 10 percent of federal legislation required the *Bundesrat's* approval.

However, the power of examining bills in the *Bundesrat* has significantly increased in recent years. In 1993 around 60 percent of federal legislation required the upper house's examination. The attitude of the *Bundesrat* toward PWR legislation depends on the political background of its majority party. In the early 1990s, the opposition *SPD* (the Social Democrat Party) and the *Bündnis'90/Grünen* (the Green Party) secured a two-thirds majority in the *Bundesrat*, but they did not hinder much the *CDU/CSU-FDP* (a coalition of Christian Democratic Union, Christian Social Union, and Free Democratic Party) government to enact PWR legislations such as its important Avoidance and Recovery of Packaging waste Ordinance (ARPWO) in 1991.

However, after these two major political coalition parties swapped their positions after the 1998 election, the new government, composed of the "red-green" coalition of the SPD and the Greens, has quite often faced strong opposition from the majority of the right conservatives in the *Bundesrat*. For example, in early 2001 the Federal Environment Ministry and the Federal Ministry of Economics and Technology proposed a mandatory deposit law on drink cans, disposable glass bottles, and disposable plastic (PET) bottles. The *Bundesrat* rejected it in July of the same year. This explains the smooth making and implementation of PWR laws and regulations during the Kohl government and the rough ride doing the same during the Schröder government.

Although Germany has a federal system, the federal executives have made delicate efforts to acquire a unified PWR strategy in the country. Under the principle of federal laws superseding state (*Länder*) laws in the waste management area, the *Bundestag* enacted the 1977 German Waste Law. Government and industry had been debating the promotion of bottle refilling. But the former did not concede the latter the right to use more nonrefillable bottles. Besides, in response to the industry's failure to increase the recycling rate of nonrefillable bottles, the government established a deposit ordinance for one-way PET bottles in 1989. This enables a closed cycle for PET bottles that can be directly sorted out and recycled without mixing up with any other material. Since most German beverage companies use refillable bottles, other European packagers lodged complaints with the European Community (EC) Commission that the law harmed cross-border trading. Under pressure from the EC Commission, the *Bundestag* replaced the law by enacting ARPWO in 1991, introducing a packaging waste producer responsibility system and requesting beverage companies to package at least 72 percent of the volume of their products in refillable containers and, once used, to retrieve these containers. As the ordinance shifted the cost of PWR from the public sector to private industry, the latter strongly opposed the law. The government has thus implemented the law in a delicate way by only setting quotas for the industry and not resorting to penalties for failing to reach the quotas, but warning industries that mandatory deposits for all nonrefillable liquid containers would be imposed if they could not reach the stipulated refilling quotas (all except milk 72 percent, water 91 percent, carbonated soft drinks 74 percent, juices 35 percent, beer 82 percent, wine 29 percent, milk 20 percent) and the recycling targets (glass 90 percent, aluminum 90 percent, and plastic 80 percent) by January 1995. German industries and retailers negotiated with the government and tried to avoid mandatory deposits by organizing a collection and recycling network as an alternative to the usual municipal waste handling system. This response touched off the assembly of *Duales System Deutschland AG (DSD)*, retrieving and recycling packaging waste according to nonmandatory recycling targets.[13] The *DSD* acted as a private company under a monopolized license to collect, sort out, and transport waste to recyclers, cooperating with a large number of members in the industry and providing a complete collection and recycling process so as to share the cost by a large pool of companies or producers. The *DSD* sent packaging waste to its contracted recyclers and had to report its annual performance to environmental ministries for verification of its fulfillment. Producers/bottlers could directly and voluntarily enter contracts with the *DSD* to release their collection and recycling obligation and apply the Green Dot mark to their container by paying license fees; otherwise they had to collect and recycle their own packaging waste themselves. However, the operation of *DSD* was not so smooth at its inception. The monopolized business had seriously underestimated the cost for the collection, distribution, and recycling system at about \$2.1 billion per year in 1991 dollars. Soon after the initial start-up the cost was found to be about \$4.2 billion. It collected about four times the expected quantity of materials in the first year, but the Green Dot licensing fee received was sufficient to cover only 60 percent of these materials. In response to the problems, the overdue bills of the waste-recovery firms were to be converted into long-term debt by the recyclers.[14]

Meanwhile, the 1994 Container and Packaging Waste Directive (CPWD) of the European Community requested the EU member States to set up their own systems to achieve high recycling rates for all kinds of containers. The CPWD set 50 to 65 percent of waste recovery rate, 25 to 45 percent overall recycling rate, and a minimum of 15 percent for each material type recycling. These targets had to be fulfilled by 2001 (2006 for Greece, Ireland, and Portugal). Although these cross-border requirements are less demanding than the German ones, the government emphasized the importance of the national PWR development by adding Article 20a to the Basic Law (the German Constitution) in 1994, expressly committing the state to "protect the natural bases for life."[15] The Federal Government enacted a revision of the ARPWO in November 1996, requiring industries to give proof of their participation in *DSD* and strengthening the percentage quotas of PWR. However, the *Bundesrat* rejected this first revision in April 1997. The EU Commission also responded with objections. In response to the veto by the *Bundesrat*, the executive did not give up but circumvented the resistance by upholding the existing ARPWO to impose deposits on only those beverages packaging that did not meet their quotas. After long and largely legal battles in courts, the federal government finally included one-way packaging costing 25 to 50 Euro cents for mineral water, beer, and sparkling soft drinks in a CDL that took effect on January 1, 2003. Although the refillable proportion in the wine and fruit juice sectors fell under the

1991 quotas, those beverages were not subject to the CDL until later when surveys showed that consumers clearly favored the deposit plan, and a deposit was imposed also on all other one-way beverage packaging nine months later on October 1, 2003.

5.7   Sweden

The Swedish executive branch is characterized by its bifurcate roles of policy formulation and implementation. The prime minister and cabinet ministers are those who initiate bills to be voted in the multiparty parliament. Due to the fact that for many years no single political party in Sweden has achieved more than 50 percent of the vote, political parties always form coalition governments and cooperate on law-making. Thus, *Riksdag* (the Swedish unicameral parliament) members usually support the cabinet ministers who are not necessarily parliamentary members . Also, it is a tradition that a large number of *Riskdag* members are enlisted at the prelegislation stage. Most *Riskdag* members including those of opposition parties have often served on commissions of inquiry or deliberation (Arter in Philip Norton ed. 1990: 125–126). There are hundreds of commissions preparing possible bills. The commission will then conclude their discussions in official reports and submit them to the responsible minister. The executive controls the commission system by issuing directives and determining the composition and the time scheduled on bill deliberations. Consecutively, the minister will invite business associations, trade unions, and specialists for further discussions. Different social opinions may also be included. By absorbing the opinion of minorities and nongovernment bodies, arguments during parliamentary voting do not arise. Under this state structure, PWR bills proposed by government and endorsed by the Parliament and the business prior to public announcement are highly likely to become law with less industrial opposition and repeal. Although Sweden has 21 counties, it does not have a federal system. Each county has a County Administrative Board executing national political goals, and the County Council appointed by the county electorate is responsible for only the public health care system. Thus, Sweden has unified PWR laws and regulations throughout the country.

Under this state structure, Sweden has a strong state capacity to motivate industry and becomes the frontier-country to develop the concept of producer responsibility. The Swedish story of producer responsibility started in 1973. For offsetting the revenue loss arising from food price suspension, the Swedish government enacted tax on beverage containers from beverage producers, bottlers, and importers.

The country abolished this tax in 1984 and soon replaced it with a tax scheme that affected only containers that were subject to a deposit. The tax lasted until 1993, when the government eradicated it and applied producer responsibility regulations for different types of container through the Eco Cycle Bill. Unlike Germany, Sweden did not have any intention to include provisions specifically for refillable beverage containers. Instead, the country has allowed more one-way beverage containers. Recycling rates were achieved by organizing *ad hoc industry-managed* collection and recycling systems first for aluminum cans and later for PET and glass bottles as well. To drive the industry and realize this, in May 1982 the Swedish Parliament approved the recycling law SFS (1982: 349 in 1982 amended in 1998) for the recycling of aluminum cans, taking effect from March 1984, promulgating that an aluminum can deposit system be introduced, and requesting a 75 percent recycling rate. This target was quickly reached and thus a higher target of 90 percent was reset. In 1997, 90.5 percent of cans were collected. The collection system and recycling coordination for aluminum cans was managed by *AB Svenska Returpack*, a nonprofit organization found in 1984 by the packaging industry, trade organizations, and the Swedish Brewers Association.

Following this success, the Swedish legislation passed SFS (1991:336 amended by SFS 1993:418) aiming at improving the recycling rate of PET-bottles. *AB Svenska Returpack-PET* as a cooperative was also founded by the industry it and shared some of its administrational and informational resources with *AB Svenska Returpack*.[16] These two independent companies cooperated with each other to administer their respective deposit and recycling systems. Having considered that the voluntary deposit system was mature enough, in 1994 the government required that industries acquire a minimum recycling rate of 90 percent for both aluminum cans and PET bottles under the 1994 Ordinance on Producers' Responsibility for Packaging (OPRP) (revised in 1997). Also, the OPRP required the glass industry to reach a recycling target of 70 percent for glass containers. This target was reached quickly by the glass collection and recycling efforts of the industry-run *Svensk GlasÅtervinning AB* in 1996.[17] Due to these arrangements, the three specially supported beverage packaging systems have achieved outstanding recycling rates in comparison with others in the country. As of June 30, 2001, the recycling rates of aluminum cans, PET, glass, and general plastic were 90 percent, 90 percent, 86 percent and 70 percent respectively. The difference between these recycling rates is obviously caused by the *Returpack*'s and *GlasÅtervinning*'s special focus on the former three rather than the latter one.

In addition to the industry's efforts, the government has been close to these industry-run organizations in ways that enable the officials to assist PWR development in various ways. First, most 33 centiliter refillable cans in Sweden must be of uniform shape and size to eliminate sorting by brand. Second, consumers have to return their empty cans back to the specially made Reverse Vending machines (RVM) that accept and crush empty cans and refund 50 öre per can.[18] The beverage producers/bottlers would collect the compressed cans that are baled and delivered to the smelters for repossessing and then delivered back to the packaging industry represented by Rexam for remanufacturing. Third, the government has arranged collection points to be densely scattered across the country. According to the Swedish Environmental Protection Agency (*Naturvårdsverket*),[19] some 7,700 recycling stations were installed in the country in 2003, serving the population with a ratio of 1:1153.

5.8   SWITZERLAND

In Switzerland, the executive (Federal Council) and the bicameral legislature (Federal Assembly) cooperate intimately with each other to pass bills (Laundy 1989: 71). Although the Federal Assembly elects and supervises Federal Councilors, the latter has an outstanding role in both arranging the legislative process and executing federal laws. The Federal Council is designed to accommodate every major parliamentary party to form a coalition government so that bills can be initiated by a stable governmental operation through compromise. Since 1959, the four major parties have been represented in the Federal Council in proportion to the number of their representation in the Federal Assembly. In addition, after working out the bills in detail at committee level, amended bills are shuttled between the two legislative chambers, that is, National Council and the Council of States, until mutual consent is made (Tsebelis and Money 1997: 189–194). PWR bills usually undergo these multisteps of compromise or trade-off before becoming a law.

Under this multipartite coalition of Federal Council, the Swiss government is able to coordinate the PWR market easily and Switzerland has become a preeminent country for recycling. Its principle of "producer responsibility" has been interpreted by an intimate government-industry relationship. The 1983 Federal Law on Protection of the Environment (revised in 1995) was a major federal law that used the principle of "producer responsibility" to govern general waste management in Switzerland. In 1990, the Swiss government enacted the Ordinance on Beverage Container Packaging (revised in 2000) for some types of drinks, covering containers for mineral water, carbonated soft-drinks, and beer. Nevertheless, the government behaved very carefully in regard to nonrefillable bottles. Like Sweden, the state rules on these one-way bottles were applied one by one, finally exempting consumers from paying mandatory deposits. Aluminum cans were the government's initial target. The government discussed with the aluminum industry about how to achieve high recycling rates. Thus, IGORA Co-operative (IGORA) was set up in 1989 and was composed of members from the drink industry and aluminum industry,[20] and it undertook the collection and recycling of consumed aluminum containers. The collection and recycling systems were supported by local authorities, retailers, the IGORA's industrial members, and recyclers. While a CDL was introduced for all refillable containers by the law, fillers and importers of nonrefillable aluminum cans were required to voluntarily pre-pay recycling fees of CHF 0.04 for each can sold in the country to IGORA, otherwise they had to develop their own collection and recycling systems in their own right, aiming at meeting the government's recycling targets. In addition to these prepaid incomes, IGORA used revenues from the sale of scrap and further financial contributions from the aluminum industry to pay CHF 1.30 per kilogram to collectors who delivered packaging waste to scrap dealers. According to the statistics of the IGORA, aluminum can recycling increased from 31 percent in 1989, to 91 percent in 2000, achieving a reputation as a world leader in this recycling field for Switzerland. Meanwhile, PET-Recycling Schweiz (PRS) for PET bottles, with 90 percent of its members coming from the beverage industry, was also founded in 1990 to establish a national PET-bottle collection and recycling system. Since 1991 the PRS has also not introduced any CDL for empty bottle collection (Migros, a supermarket chain, was the only retailer in Switzerland that requested a deposit on one-way PET bottles then) and the only financial instrument maintaining the operation was dependent on a recycling fee of 10 Rappen (Swiss centime). Owing to a fast growth of collection volume and a better efficiency of handling, the recycling fee was reduced to 5 Rappen in 1996 and was further reduced to 4 Rappen in 2000. Under the management by and cooperation between Swiss beverage industries (producers, importers, bottlers, and retailers),[21] the PET-bottle recycling rate had risen to 82 percent over a ten-year period, one of the highest rates worldwide. Although the recycling rate target for PET-bottles had not yet been obtained, the actual recycling rate was so close to the target that the government did not impose mandatory deposit and even lowered the residual waste amount limit, encouraging the industry to make further headway. Likewise, the glass PWR is also financially administered by a monopolized, private organization, *VetroSwiss.* A disposal charge of CHF 0.02 to 0.06 per bottle is prepaid on a statutory basis. A color-sorted collection system is adopted. Waste glass bottles are collected for recycling or recovery with the help of communal authorities. Thus, the cost can be further shared by local governments. With these state-societal joint efforts, the return rate of waste glass bottles in Switzerland rose to 94 percent in 2002.

Other than the self-managed industrial networks, the success of acquiring high PWR in Switzerland is also to the credit of the coordination and side works provided by the government that is always in close contact with IGORA and PRS. Although the industry rather than the government is responsible for the recycling arrangement for municipal and business waste collection and recycling under the principle of "producer responsibility," city authorities generally took the initiative to contract out these tasks to local companies. Although the industry is the one that knows how to collect, process, and eliminate its own waste, the job of the government is still indispensable. Other than setting recycling goals for industries, the government's strategy to motivate its industries is to threaten them with a mandatory deposit-refund system wherein the industry would bear a much higher cost. This threat pushed the industry to evolve recycling systems. Further the concept of "producer responsibility" in Switzerland is not to ask producers to pay all the additional cost. The industry was not the only one to shoulder the cost of recycling, households too had to pay a "ground fee" for waste recycling service and "bin-liner fee" (special plastic bags for their residual waste) for municipal waste management and collection. The "bin-liner fee" encouraged citizens to return their empty bottles to retail shops to claim deposit rather than dispose them from home paying a "bin-liner fee." Moreover, in addition to the large network of retail shops and privately owned collection centers, the government also established many staffed collection points and container parks to make sure that an easily accessible infrastructure for consumers to return their consumed bottles was available. For example, according to IGORA in 2003, 47,000 PET collection receivers were installed mostly near retail shops, train stations, kiosks, or restaurants, and 18,000 PET bottle collection points (each serving 405 people) were appointed or erected by local governments.

In short, Swiss government, like the German and Swedish governments, could achieve rapidly growing recycling rates by concentrating on individual industries at the early stage and acting as a complementary provider with services and facilities for the monopolized market actors.

5.9   Conclusions

This chapter has compared six countries and investigated whether or not their institutional structures in which national and local governments maneuver enforce what they can enact, and that the manner in which they have worked with industries and other social groups have led to divergent results. First, it was examined whether their legislative norms and government structures restrict or facilitate governments to enact PWR laws and regulations at national and local levels. The comparison has revealed how governments were obstructed or supported by their respective legislative norms and government structures in making PWR policies and in requesting their industries to increase PWR rates. These comparisons addressed the impact the state-societal structures (made of national executives, national legislatures, local governments, industries, and other social groups) had on the making and implementation of PWR laws and regulations, and their outcomes, in the United States, the United Kingdom, Australia, Germany, Sweden, and Switzerland (table 5.2).

Given these comparisons, the findings suggest that, first, the type of legislative norm ("discord" or "compromising" legislature) and the kind of government structure ("majoritarian" or "coalition" executive) prove vital to the outcome of the respective national and local PWR status. While the "majoritarian" executives in the three Anglo-Saxon countries were prevented from efficiently making PWR laws and regulations by the opposition of "minority" parties and the industry through their "discord" legislature that allows more law-making autonomy for nongovernment legislators and local governments (worst in the United States, moderate in Australia, and fair in the United Kingdom), the "coalition" executives in the three European countries—especially prior to the legislative stage—resorted to compromise and/or collaboration among different political parties and local authorities for their PWR developments (much in Switzerland, more in Sweden, and less in Germany). These diversified state structures explain the incoherent and unstable PWR laws and regulations in the three Anglo-Saxon countries and the consistent and steady PWR laws and regulations in the European countries.

Second, under the distinct state structures, government-industry cooperation in PWR programs took effect in different ways. While the incoherent and unstable PWR laws and regulations in the three Anglo-Saxon countries discouraged individual companies from investing in their own PWR programs, the consistent and steady PWR laws and regulations in the European countries endorsed cooperation of producers to set up their own PWR system.

Table 5.2 Impacts of State-Societal Structures on PWR Rates

	The United States	The United Kingdom	Australia	Germany	Sweden	Switzerland
Main PWR Bills	By Majoritarian Government	By Majoritarian Government	By Majoritarian Government	By Coalition Government	By Coalition Government	By Coalition Government
Opposing Private Bills	Often	Sometimes	Sometimes	Seldom	Seldom	Seldom
Legislature's Attitude	Opposition or Discord	Opposition or Discord	Opposition or Discord	Compromise or Collaboration	Compromise or Collaboration	Compromise or Collaboration
National PWR Laws and Regulations	Subtitle D of Resources Conservation and Recovery Act (1976)	Producer Responsibility Obligations (Packaging Waste) (1997)	National Packaging Covenant (1999) NEPM (1999)	Avoidance and Recovery of Packaging Waste Ordinance (1991/1998)	SFS 349 (1982/1994) SFS 1154 (1993) SFS 336 (1993)	Statutory Order on Container for Beer and Soft Drinks (1990/2000)
Local PWR Laws & Regulations	Different from Federal Laws (Ten CDL States)	Devolved from National Laws	Different from Federal Laws (One CDL State)	Overridden by Federal Laws	Overridden by National Laws	Overridden by Federal Laws

Continued

Table 5.2 Continued

	The United States	The United Kingdom	Australia	Germany	Sweden	Switzerland
Ad hoc and/or monopolized recycling system for limited types of packaging waste	Not Concentrating on any type of beverage bottles except those CDL States	Not concentrating on any type of beverage bottles	Not Concentrating on any type of beverage bottles except the CDL State	1989 regulations requiring one-way PET bottles 1991 monopolized *DSD* for most beverage bottles	1984 *Svenska Returpack* for aluminum cans 1991 *Svenska Returpack-PET* for PET bottles 1996 *Svensk GlasÅtervinning* for glass bottles	1989 *IGORA* for aluminum cans 1990 *PRS* for PET bottles
State-Societal Institutionalized Dialogue	Nil	Since 2002 Via ACP	Since 1999 Via NPC Council	Since 1991 Via *DSD*	Since 1984 Via *Returpack*	Since 1989 Via *IGORA*
National PWR Rates (in the 1990s)	Laggard and Declining	Laggard but Ascending	Laggard but Ascending	Leading and Ascending	Leading and Ascending	Leading and Ascending

On the one hand, as the industry was unwilling to make PWR investment in the Anglo-Saxon countries, it was not realistic to impose PWR quotas/targets on the industries. The United Kingdom has, however, since 1994, resorted to introducing the Producer Responsibility Obligations (Packaging Waste) Regulations to comply with the EU laws, but the government later introduced trading PRNs to take some pressure off the industry. Nevertheless, due to the legislative restrictions in the state structure and the general dirigiste PWR policies, the performance of PWR policy in these Anglo-Saxon countries is often impaired. As a result, among the sampled Anglo-Saxon countries, the American PWR development lags behind the European countries the most. The federal government is unable to manage the PWR policy issue that has been contingent upon the result of the struggle between proponents and opponents in the legislature. The British PWR development also lags but it has improved since the introduction of the Producer Responsibility Obligations (Packaging Waste) Regulations in 1997 and institutionalized government-industry dialogue through ACP in 2002. But, its recent parliamentary reform in the late 1990s will call its unified PWR policy into question in the near future. The laggard Australian PWR development has also been improved since the resumption of national involvement through the ANZECC in 1991 overriding the state governments, and since the beginning of the public-private participation in the NPC Council and the enactment of NPC/NEPM by the Commonwealth in 1999 allowing industries to devise their own alternatives. Its future PWR development will be a matter of whether or not the federal government can overpower the federal upper house and local governments/territories through the international treaties, the newly institutionalized settlement of conflicts with industrial/social groups for PWR law-making, and fulfill its own commitments.

On the other hand, the cooperative relationship among political parties in the three European coalition governments has created long-term and stable regulatory systems, giving industries the confidence to commit to their future success. The European examples show that the governments that coordinated the market for the industry to smoothly manage the collection and recycling system produce progressively better results. Although industries' antipathy toward CDL existed also in the sampled European countries, the coalition government structure did not allow the industry much room for lobby and repeal. In addition, although the European governments used CDL to threaten their industries, the coalition governments showed consideration to the industry's capabilities by postponing/suspending the CDL and/or imposing less incumbent/financial pressure. Further, the coalition governments regulated their industries, step by step, carefully. As the governments entrusted their industries with the task of organizing ad hoc and/or monopolized recycling system for *limited* types of packaging waste, the cost was reduced to a minimum and thus easily shared by the whole industry. This pilot project was first seen in Germany when the government chose PET in 1989 as the first item to be directly sorted out and recycled in a closed system without getting in touch with any other material. Sweden and Switzerland also have similar arrangements. They permit industries of a single packaging type to monopolize the collection of their own containers. The governments did not allow industries to extend their PWR systems to other packaging types until the pilot industry-run programs had made considerable progress. Moreover, the governments have made various supportive efforts to ease both resource and financial pressure on the industry, making efforts such as erecting more collecting points and/or introducing "bin-liner fee" to encourage consumers to return bottles more to collection centers than to household bins, thus promoting the industry-run collecting and recycling systems and reducing conflicts of interest to a minimum.

These explain that industries in these three European countries were more willing to suffer short-term, affordable loss through their collection and recycling system, comply with product design required by governments, and even reduce handling fees when their collection and recycling system became more efficient. In contrast, their Anglo-Saxon counterparts have refused to adopt or increase government requirements, for example, using accredited PET-content for their bottles, but embracing more marketable but nonrecyclable product designs and materials, and having a penchant for landfills that often offer a cheaper solution. This behavior confined the possible PWR policies to financial incentives and government-paid recycling handling systems. As a result, the distribution of financial incentives was always a contentious issue due to the confrontational positions taken by governments vis-à-vis the industries conditioned by their particular state structures and liberal PWR policies.

These results, especially in the United States, are consistent with the findings of Wood (2001: 274) that dirigiste mechanisms of policymaking by "designing a set of sanctions and incentives and imposing these patterns of behavior on firms" is ineffective. Yet, Australia has improved its laggard PWR development by increasing its federal policymaking powers through ANZECC since 1991 and its policy-implementing powers by installing an institutionalized public-private dialogue mechanism in 1999; the United Kingdom has also made a similar modification to its public-private communication by restructuring its ACP but only since 2002. By contrast, government-industry institutionalized settlements of conflicts of interest with industries at the PWR policy formulating as well as implementing stages have been consistent in the three European countries over the past two decades. Therefore, the noticeably different policy outcomes between the Anglo-Saxon and European countries are consistent with the basic assertion of the "broad" institutional study (e.g., Hall and Soskice 2001; Wood 2001; Weiss 2003). In addition to the numerous studies on whether the type and quality of PWR programs matter to PWR development, the evidence in these case studies has shown that effective PWR policies are also dependent on whether their policies have adapted to their respective institutional structures, especially in terms of state-societal structures.

After studying the impact of different state structures and their relation to society on environmental-technology development, two environmentally friendly auto policies, both of which contain heavy regulations and government-business cooperative programs but each employs different tactics to manage government-business relationship, will be further examined in the next chapter.

6

Varieties of State-Societal Relationship: Environmentally Friendly Auto Technology Advancement in the United States and Japan

6.1 Introduction

According to an IEA World Energy Outlook survey, consumption of oil by transport in 2000 increased from around 44 percent in 1990 to nearly 55 percent of total production in OECD countries.[1] The transport sector is also responsible for most greenhouse gas emissions. These energy security and environmental pollution issues are consequently placing national governments and automaking industries under social and political pressure to improve vehicular energy efficiency and search for cleaner energy alternatives for the transport sector. The rapid progress of environmentally friendly auto technology (EFAT) over the last decade has thus led to the transition of the diesel-dominated car industry to an EFAT market economy.[2] However, technological breakthroughs have not been adequate to diffuse the technology and enlarge the vehicle market. If EFAT's price and performance were sufficiently competitive, markets would be created by the so-called invisible hand; but, in terms of actual political economy, car manufacturers have come across many difficulties that cannot be resolved by market players themselves, that is, neither by automakers nor by auto purchasers. These problems range from consumers' perceptions and needs, standards and custom barriers, insufficient global infrastructure, and even the confusion of there being too many unknown and unproven prospective clean-fuel vehicle solutions. The future of EFATs is yet to be determined.[3]

Of these developmental problems, the most immediate puzzle is who should invest the enormous amount of public capital and make the technological choice required by the automotive development: the government or the industry? The very nature of extremely large, incremental capital investment implies that once a market player has committed to developing a certain type of environmentally friendly auto, the required investment becomes costly and time consuming. The investment in environmentally friendly car production also includes an astronomically costly infrastructure such as clean fuel or electricity filing stations, promotion, training, and maintenance. Moreover, it is meaninglessly absurd and unjustifiable for an individual company or consortium to propose to build public infrastructure that will also be enjoyed by its competitors. Thus, automakers and other investors are highly cautious of involving themselves in any particular clean-fuel design and/or other complementary stakes. Not only is the product noncompetitive, but also the absence of a higher degree of government involvement has seriously repressed environmentally friendly auto market access. This phenomenon of insufficient state involvement was fairly common before the 1990s.

For the above reasons, in addition to the responsibilities connected with the threat of the scarcity of oil, and CO_2 emission being one of the major causes of global warming,[4] governments cannot pass the buck to market players but must make some critical decisions and get involved in this transition. Starting from the early 1990s, policymakers in these industrialized countries have addressed a number of alternatives in an attempt to speed up improvement of EFAT through government involvement. In this regard, many countries have intensively participated in EFAT development process, including even those that ran the most liberal market economies such as the United States (the groundbreaking American Partnership for a New Generation of Vehicle (PNGV) public-private cooperation program), as well as those that ran well-coordinated market economies such as Japan (Advanced Clean Energy (ACE) Vehicles government-industry joint-force program).[5] These joint efforts were operated by governmental and industrial representatives, working closely to seek technological innovation toward their common program goals, that is, an ambitious target of largely improving automotive fuel efficiency and switching to cleaner fuels.

However, the issue of government involvement in the transition from EFAT invention to diffusion is far more complicated than ever. Though involved in fixing market and systemic failure, setting automotive environmental and safety standards, building public clean-fuel supply infrastructure, encouraging and protecting investment, exchanging information and resolving many other issues, the performance of these jobs is based on the quality of government-industry relationships. Policymaking, administration and progress monitoring involve "easy-to-err" bureaucratic observation and judgment, yet an effective interactive relationship between government and industry may broaden policy capacities and help minimize government failures. Showing how both government and industry are aware of the transformative problem, and effectively carry out remedies through institutional approaches and patterns, has recently become a central issue in policy dissection.

The purpose of this chapter is to demonstrate in detail the impact of institutional structures on the quality of government-industry partnership programs. In doing so, the institutional contents of the American and Japanese EFAT government-industry partnership programs will be compared and scrutinized, arguing that quick, neutral, and shrewd bureaucratic efforts, rather than relatively slow, political, and holistic parliamentary or congressional decisions, matter to the more successful partnership outcomes.[6]

The impact of institutions on the quality of state-private partnerships will be discussed followed by EFAT development case studies in the United States and Japan. The U.S. case study will be scrutinized first, examining the context for the emergence of PNGV and the formation of its institutional arrangement. By using an institutional approach, empirical evidence explains why the EFAT public-private cooperative programs over the 1990s did not work and led to an early termination of the program. The next section of the chapter analyzes the corresponding case of Japan. It illustrates that its old form of Japanese state power did not adapt to the new era in the 1990s and how the government modified this old form of coordination, finally making fast progress in EFAT development. The case demonstrates in detail how, under the specific Japanese institutional structure, prudent and discerning bureaucracies manage resources for the innovation and acceleration of the development of EFAT. The concluding section then explores the implications of the case studies for distinguishing the forms of state involvement in the two different domestic institutional structures.

6.2   THE IMPACT OF INSTITUTIONAL STRUCTURES ON THE QUALITY OF PUBLIC-PRIVATE PARTNERSHIPS

In immature technological and market conditions, bureaucratic decisions and practices on what and how to develop are based on personal expertise, management technique, and market flair, though the business sector is consulted. They are preoccupied with making optimum decisions that public and private parties can find satisfactory. However, synergistic public-private relationships to date have not been thoroughly theorized. One of the most inspiring attempts made in recent years is a publication edited by Peter Evans (1997). His argument was based on the articles—by five different scholars—in the book, holding the view that public-private "synergy usually combines 'complementarity' (i.e., government may complement deficiencies of the private sector) with 'embeddedness' (i.e., government should embed in the private sector for better carrying out the 'complementarity') and is most easily fostered in societies characterized by egalitarian social structures and robust, coherent bureaucracies." Another inspiring academic breakthrough is the concept of "governed independence" (GI) by Weiss (1998) as discussed in chapter 2. The four major types of GI, that is, "disciplined support," "public risk absorption," "private-sector governance," and "public-private innovation alliances" are mutually agreed modes of government-industry cooperation. Further literature on public-private partnership can be found in an OECD publication (2001:18), *Local Partnerships for Better Governance*. This review points out that "Partnerships are sometimes compared to a 'black box': inputs and outputs are visible, but the mechanisms enabling the transformation from input to output are not." This OECD's study (2001: 127–128) proposes the following four techniques: (1) "Make policy goals consist at central level"; (2) "Adapt the strategic framework for the partnership to the needs of the partners"; (3) "Strengthen the accountability of partnerships"; and (4) "Provide flexibility in the management of public programmes" as the recommended strategy's four "axes." These complement and strengthen each other and can generate improvements in the performance of public-private partnerships.

Difficulties emerge, however, when we use concepts in the above literature to compare the U.S. Partnership for a New Generation of Vehicles (PNGV) and Japanese public-private cooperative automotive programs in the 1990s. First, during the early 1990s, given a wide variety of possible EFATs requiring *different* and *conflicting* public infrastructure, for example, liquefied petroleum gas (LPG), compressed natural gas (CNG), or electric battery refilling/recharge networks, how could governments *complement* the enormous and incompatible public goods of egalitarian social structures as discussed by Evans? In addition, "complementarity" equipped with "embeddedness" works well by rounding off the private actor's deficiency with governmental efforts only when they have the same goals. Norms of trust and social capital from intimate partnership interaction are built up only when they have common interests. What if they have different goals and/or distinct interests, for instance, disagreement between environmental improvement and corporate profitability?

Second, many profitable automakers in the United States and Japan do not need government financial support or government equity involvement. Also, the enormous extent and international nature of the auto market make it difficult for governments to absorb the risk of adverse deviations in investment. Moreover, the auto industry has always had a serious conflict of interest with government environmental goals. This has made public-private negotiation very difficult, for example, the deep divisions in the United States gave rise to lawsuits against the auto industry in the early 1970s. Thus, without common goals and needs for interdependence between government and the auto industry, GI is inapplicable to early EFA development. It would appear that the structure of the American PNGV resembles the "public-private innovation alliances" form of GI, but the conflicting goals between the U.S. government and Detroit's Big Three had transformed it into a nongenuine alliance. Nor have the Japanese EFA public-private cooperative programs operated in any of the other three forms of Weiss's GI governance models since neither "disciplined support" nor "public risk absorption" nor "private-sector governance" have been employed. In addition, later successful hybrid technologies had not been included in the early public-private cooperative programs at all. In fact, some strong Japanese automakers such as Toyota or Honda are no longer heavily dependent on government for technological advancement.

Third, the management of the PNGV program had apparently incorporated the above-mentioned four "axes" of the OECD's proposal from the outset, and in that case why has the pace of American, environmentally friendly auto innovation substantially lagged behind its Japanese counterpart? Is a fifth "axis" involved in the equation?

The previous two chapters have proclaimed the significance of Japanese government-industry relationships in the car industry. Wallace (1995: 235) points out that "this (Japanese) attitude to innovation led to an increasing gap between the quality and performance of Japanese cars and their American rivals." He then suggests that high-quality, honest public-private dialogue is crucial to technological responses to vehicle pollution. Also, Sperling (2000: 8) indicates that "other smaller automakers from elsewhere (Japan and Germany) made faster progress in advancing the commercialization of targeted technologies than the U.S. automakers."

In fact, around the end of the last century the American Big Three automakers under the PNGV program could, unfortunately, exhibit only diesel-electric hybrid prototypes that failed to fulfill the high emission standards in California. Some of them could only fulfill a loose Emission Vehicle mandate after the PNGV discontinuation, for example, Ford's Focus PZEV in the spring of 2003.[7] GM and Daimler filed a suit against the mandate in the federal court. As a result, they won an injunction postponing the mandate until 2005. In contrast, two purely Japanese-owned automakers, Toyota and Honda, have since begun to produce massive numbers of gasoline-electric hybrid vehicles in full compliance with the same Californian mandate. At length, the Californian Air Resources Board (CARB) certified the Japanese hybrids as fulfilling its stringent Advanced Technology Partial Zero Emission Vehicles (AT-PZEV) standards.[8]

These divergent EFAT development outcomes between the United States and Japan can be attributed to many factors. Gas price appears to be a major factor for rapid sales because Japan has nearly two times higher retail gas price than that of the United States,[9] but it cannot be a factor for their rapid commercialization and market launch in the mid-1990s. Subsidization is not a major causal factor of technological success because the Japanese government has invested much smaller amounts of money than the U.S. government through R&D subsidization (US$35 million versus US$260 million). Rather, the institutional difference between the two countries is a more convincing explanatory variable. Sperling (2000) showed that the annual reports published by The National Research Council from 1994 to 2000 focusing on the management of PNGV could not address all possible factors in the making of a growing EFAT market.[10] Thus, looking only at the cooperation program is insufficient. A more comprehensive study of the variables directly interacting with the cooperation program would be more appropriate for a better understanding of why similar public-private cooperative programs result in different outcomes. This indicates that the *type* of political economy, in addition to public-private cooperation itself, can also be the factor. In this regard, the institutional structure of public-private cooperation may be a key issue in the elucidation of why some industrial transformations from conventional to environmental technologies are more effective than others under similar public-private cooperative programs.

Therefore, there are some questions to be answered: Why did the American automotive industry behave less enthusiastically and why was it less committed to a public-private partnership, whereas the Japanese counterparts were more successful through working in partnership? What are the actual roles and responsibilities of government in successful cooperation? What is the actual "division of labor" between the government and the car industry that enhance better environmental innovation? How did successful cooperation influence EFA production decisions and consumer choices? What are the differences of institutional structure in these two countries?

This chapter takes an institutional approach in dealing with these questions through a comparison of the American and Japanese government-industry partnerships in the auto industry. As a matter of fact, the American PNGV public-private cooperative program symbolizes a major deviation from the traditionally antagonistic relationship between government and the industry in the automaking industry, imitating the state-oriented policy in some special areas, such as semiconductors, flat panel displays, and defense-related industries, but isolating it from most other regulated or deregulated consumer industries in the country. This state-industry partnership was created in reference to the Japanese government's heavy support of research and development to the industry. However, their different political and business structures and practices remain unchanged. Thus, this chapter will explore how different structures of political institutions and industrial organizations, and intensities of government involvement, may shape public-private relationships by a comparative study of EFAT development over the last 20 years. A broad institutional approach, that is, political and business structural context, in which legislators, government officials, industry's representatives manage the public-private cooperative programs, is employed.

6.3   Comparative Case Studies between the United States and Japan

The groups of case study in the last two chapters have demonstrated the significance of (1) laws and regulations and (2) state-societal structures. But government-industry partnership programs in the auto industry in the United States and Japan that possessed these two institutional attributes produced distinctively different policy outcomes. Thus they are worthy of being compared and analyzed in greater detail. The objective of the two case studies of this chapter is to investigate in depth the different institutional arrangements between the United States and Japan, attempting to explain why the latter's is more effective to the innovation and diffusion of EFAT over the last decade. This comparison focuses on five institutional layers greatly impacted upon by the public-private cooperative programs in the two countries, that is, corporate governance norms, parliamentary or congressional structure, auto industry structure, state involvement in program review, and state power in solving program problems. The first institutional layer may reveal that different corporate governance norms create different corporate behavior. For example, "companies with access to finance that is not entirely dependent on publicly available financial data or current returns" may have different corporate behavior from those financed by investors who emphasize "current earnings and the price of their shares on equity markets" (Hall and Soskice 1998: 22–29). This influences industry's long-term commitment to the public-private cooperative program operations. The second institutional layer may reveal that different degrees of political support may lead to different degrees of progress in policymaking and implementation. As Pempel and Muramatsu (1995) and Weiss (1998) note, capabilities of government officials can be impeded or improved upon, subject to whether political leaders share a growth agenda. The third institutional layer may show that the organizational arrangement of the auto industry facilitates or obstructs the cooperation between state and industry. Industry associations can be highly integrated and actively participate in the making and implementation of policy (Weiss 1998: 60). The fourth and fifth institutional layers may expose how the public-private cooperative programs are reviewed and adjusted respectively. In carrying out industrial-technological transformation, whether or not competition occurs "in national efforts to identify and define technological trajectories" is important (Weiss 1998: 67). Are legislators or government officials or industry representatives able to identify and define technological trajectories of the EFA programs for creating a better competition environment in the cases? In general, these five institutional layers may influence the pattern and strategy of the involvement of government in public-private cooperation (rather than only the kind of laws and regulations imposed on the industry) and the involvement and developmental direction of automakers, and the choice of car consumers.

Comparison is made between the United States and Japan, excluding the United Kingdom and Germany. This arrangement for comparison is due not only to the former two countries being much larger in production size than the latter two,[11] but it is also due to the possibility of obtaining a more comparable specification. (This is despite the fact that Germany has already reached a 3L per 100-km target, equivalent to 80 mpg since 1997, its deficiency of vehicle size requirements has involved more vehicles of other types than the United States and Japan). These two countries are studied also because they already have conspicuously different outcomes. The PNGV program in the United States was declared dead by Energy Secretary Spencer Abraham after consuming $1.5 billion of government investment and targeted products were never produced, while those in Japan resulted in substantial sales in the market without spending taxpayers' considerable money (table 6.1).

I may assemble those components to put forward propositions that, notwithstanding the similar public-private cooperative programs, the institutional attributes of the United States do not enable more effective public-private cooperation than those of Japan in the EPA technology sector. Rather, strong state capacity through political support and bureaucratic coordination may facilitate public-private cooperation. Yet, when common goals and interests between government and industry do not exist, it is not worth staking resources, effort, and time

Table 6.1   Outcomes of Partnership Programs between Japan and the United States

	Japan	The United States
Major Government-Industry Partnership Programs since the early 1990s	High Performance Batteries for EVs (1992–2001)	U.S. Advanced Battery Consortium (1991–2000)
	Polymer Electrolyte Fuel Cells (1992–2000)	Partnership for a New Generation of Vehicles (1993–2002)
	Advanced Clean Energy Vehicles (1997–2003)	
Total budget per annum	US$ 35 million	US$ 260 million
Results of the Partnership Programs in the early 2000s	The new cars emit 42% less CO_2, 90% less CO, HC, and NOx than conventional cars and met all mandates. 150,000 cars sold in Japan and 40,000 cars exported to the United States.	Neither the 80 mpg target nor the California mandate could be met by the U.S. automakers. Prototypes were never produced and the PNGV was terminated before expiration.

Note: After the termination of the Advanced Clean Energy Vehicles in 2003, the sales of the Japanese EFAs kept growing rapidly. For example, by the end of March 2005, Toyota and Honda respectively sold 360,000 and 100,000 hybrid vehicles worldwide cumulatively.
Source: Toyota, Honda and U.S. Department of Energy.

on all of them, but the government should concentrate upon the most potential individual technologies/automakers without causing others loss or suffering. For example, when the Ministry of International Trade and Industry (MITI, renamed as METI in 2001) tried to force automakers to merge in the 1960s, the industry refused to cooperate.[12] This made MITI change its approach. The "strong" state capacity herein refers to an authoritative but patient, discerning, and "catalytic" bureaucracy, providing tailor-made institutional infrastructure favorable to individual automakers, rather than to the whole auto industry, without forcing automakers to sacrifice profits to achieve any technological targets on time. With the concentration of support in the most marketable technologies, they will become dominant so that *complementarity* of public goods and access to market niches can then be realized. In turn, there will be greater competition for market share from others.

6.4   The Evolution and Demise of PNGV

Since the 1965 Clean Air Act and the 1967 Air Quality Act, the federal agency, Department of Health, Education, and Welfare, had established automotive emissions standards and required states to submit a plan to meet their local Ambient Air Quality Standards (AAQS). However, the States were generally unwilling to enact such a legislation that would increase costs for industries that might then be forced to move to other states with less demanding legislation. Between 1967 and December 1970, only 21 States produced plans to the federal government, and none of them was approved. As the industry lobbied heavily with the state governments, it resulted in a confrontation between federal government and the auto industry. In January 1969, the Justice Department sued the Big Three automakers under the civil antitrust law for refusing to incorporate air-quality enhancers and claimed that the auto industry was to shun the development of emissions control devices that the 1965 Act made obligatory. The court imposed no penalty but issued a consent decree to prevent the auto industry from defying the request of developing air pollution control technology, prohibiting the automakers from exchanging information on emission control technology and the joint forces from writing to government regulators.[13] In 1970, the federal government enacted a new Clean Air Act, setting National Ambient Air Quality Standards (NAAQS) and requesting the States to formulate a State Implementation Plan (SIP) in nine months. However, the then newly established federal agency, the Environmental Protection Agency (EPA), lacked the power to force the States to implement the federal regulations, leading to a series of litigations.

In the early 1980s, having suffered from higher gasoline prices caused by the second OPEC oil price shock, U.S. consumers were inclined to purchase Japanese cars that generally had a lower fuel consumption rate than that of American ones. This imbalance of trade triggered the U.S. government to intervene and protect its national assets. This resulted in political settlements through the U.S.-Japan Auto Voluntary Export Restraints (VERs) during the 1980s.[14] Nevertheless, Japanese automakers tried to circumvent the VERs by increasing the quality of their vehicle production and setting up production plants in the territory of the United States, attempting to increase their profitability and to receive local treatment.[15] On the American end, the federal government strived to improve and increase the fuel efficiency of American cars and had to fortify its energy security needs along with striving to abide by its responsibility to reduce emission levels, particularly the greenhouse goals of the 1992 Rio de Janeiro Earth Summit treaty. Yet, the bitter experiences of the previous litigations taught the federal government that confrontation is not a solution. Therefore, the leftist Democrats led by the Clinton administration in 1993 tried a more cooperative approach. Public-private partnership then became the agenda to discuss whether a Japanese-style government-industry cooperative relationship, as opposed to confrontational relations, could be feasible in market-prioritized economies. At that moment both the auto industry and federal government had their reasons for entering such a partnership. A solution of this kind had become such a necessity that the government was emboldened to circumvent the previous problems of regulatory argument and confrontation between the government and industry, and to increase the competitiveness of the American auto industry that was then at a low ebb—while the auto producers might avoid the cost of fulfilling unrealistic environmental regulations, especially the prolonged pressure of fulfilling mandates of Corporate Average Fuel Economy (CAFE) standards and the Zero Emission Vehicle (ZEV) mandate that has been recently adopted in California (later followed by New York, Vermont, Massachusetts, and other American States)[16]—and sought to reduce the tremendous R&D investment and to acquire a more stable investment environment. Therefore, in the early 1990s, the Federal Council for Science, Engineering and Technology (FCCSET) formulated the Advanced Manufacturing Technology Initiative, trying to speed up the development of advanced technology for current production. Meanwhile, the national laboratories, which had just recovered from the cold war and had reduced the capacity of their weapons research, were ready to shift their expertise and facilities to national auto technical problems. These events provided the country with the prerequisites for a government-industry partnership program. Mathew Wald, a senior government official, said "we are trying to change the relationship of government and industry . . . We are trying to replace lawyers with engineers."[17]

In 1991, the government and the Detroit's Big Three-General Motors, Ford, and Chrysler (now Daimler Chrysler since 1998)—joined and formed the U.S. Advanced Battery Consortium (USABC). Following this prelude, despite the disappointing memory of the inadequate performance and high cost of electric vehicles co-researched between the industry and the U.S. Department of Energy and the Electric Power Research Institute over the last few years, the Clinton administration entered a much larger and significant agreement with the United States Council for Automotive Research (USCAR) who represented the Detroit's Big Three as a joint entity. The PNGV program, under the agreement entered into in September 1993, had a target to achieve three times more fuel efficiency than the present value, that is, a 80-mile-per-gallon typical midsize car, in ten years, without compromise of power output, size, cost, emissions, and safety in comparison with the specifications of a 1995 Chrysler Concorde, Ford Taurus, or Chevrolet Lumina.[18]

PNGV was a groundbreaking and unprecedented government-industry cooperation in the American auto industry for the national interest (President Clinton compared it to the Apollo project sending astronauts on the moon), offering a large sum of federal funding and research efforts of seven federal agencies and twenty federal laboratories and giving the Detroit's Big Three a special exemption from antitrust laws.[19] Around $250 million a year or totally around $1,500 million was budgeted and shared between the federal government and the industry with a schedule to finalize technological details by 1997, to produce a prototype by 2000, and to put them on production line by 2004. At the outset, an Operational Steering Group (OSG) and a Technical Task Force (TTF) formed by representatives from both government and industry was responsible for the coordination task. While the PNGV policy was managed by the OSG, the PNGV daily operation was carried out by the TTF. A representative from the Department of Commerce (DOC) was usually appointed chairman of the OSG and the TTF.[20] A representative from the Department of Energy (DOE) was to vice chair the TTF.

However, signs of a technological laggard of the PNGV were seen at the Tokyo Auto Show in October 1997. While the Detroit's Big Three were promoting their EFA concepts and talking enthusiastically about their preparation of prototypes in the coming few years, the public attention however focused on Toyota who announced Prius, the world's first gas-electric hybrid model that would be available for sale in December of the same year. This shocked the American auto industry that day because the Japanese model was far more environmentally advanced and technically feasible than its American counterparts. More strikingly, Toyota's technology was already mature enough for purchase. A New York-based vehicular engineering consultant Victor Wouk commented that "while we were talking about hybrids, the Japanese were building one." Four-and-a-half years later in June 2001, when the number of Prius had exceeded 50,000 in domestic sales, Toyota further introduced the Estima Hybrid Minivan using the same hybrid technology to be introduced to the market. Again in August, Toyota launched the luxury Crown, furnished with the hybrid system.

In contrast, it was not until March 2000, that the three Detroit automakers submitted their prototypes—Ford with its Prodigy, GM its Precept, and Daimler-Chrysler its ESX 2 (later its Jim Holden)—to the national administration in Washington by way of declaring their fulfillment of the PNGV agreement. Notwithstanding the submission of these prototypes, their production and commercialization are an uncertainty. These prototypes have remained in their concept stage for years. In the same month of the year, the General Accounting Office concluded that "although PNGV has made progress toward building production prototypes that meet many of the PNGV objectives, at this point it does not appear likely that such a car will be manufactured and sold to consumers."[21] In fact, the technical specification of the diesel (not gas) engines of these PNGV hybrids could not fulfill the emission standards of Californian laws. The U.S.-based automakers (except Ford who had been trying very hard to meet (part of) the requirements by developing two new models, the Focus PZEV and the Escape Hybrid) sued the Californian regulators, trying to release themselves from the mandate.[22] In an attempt to break the regulatory deadlock, the government was obliged to compromise with the U.S. automakers by lowering the standards. The Partial Zero Emission Vehicle (PZEV) was thus formulated. Spieling (2000: 14) commented that "the diesel-electric hybrid choice was the result of conservative interpretation of PNGV affordability, performance, and emissions goals, and a reluctance to re-open the discussion about scheduling and goals." On the other end of the production lines, the Big Three of Detroit kept churning out massive gas guzzling sport utility vehicles (SUVs), resulting in more inefficient vehicles than ever. Although survey results might indicate that American consumers are interested in protecting the environment and recovering the air quality, the prototypes of medium-size EFAs are seemingly not powerful enough to haul all picnic and skiing gears or even a small boat during the weekend and are too expensive both with the investment prices and maintenance costs. Having just suffered from the 1990–1991 recession, the Big Three automakers were keen on making immediate profit through corporate restructuring and popular new jeep models. According to *CFO Magazine* August 2000,[23] between 1995 and 1999, U.S. vehicle sales rose just 6 percent, from 3.9 million units to 4.1 million units but revenue was up 25 percent, an increase largely attributed to the SUVs market madness. The market signals indicated that SUVs rather than EFAs would make money.

As the chances of putting the PNGV's prototypes on production by 2004 became slim (and the recommendation of the National Research Council (NRC) Peer Review to restructure the PNGV program, a recommendation made keeping in view the diminishing prospect of PNGV's technologies as a result of the emergence of more technically and economically feasible EFAT such as hybrid or hydrogen vehicles in other countries[24]), the Bush administration thus discontinued the goals of the PNGV program and announced a FreedomCAR program on January 9, 2002, shifting the goal of public-private cooperation to fuel cell vehicles.[25] The Bush government called for a budget of $150 million for 2003 on fuel cell research. The new program has since then been principally administrated by DOE instead of DOC that had directed the PNGV program over the last eight years. FreedomCAR is a partnership entirely between DOE and USCAR. In order to better coordinate the cooperation, the Office of Transportation Technologies (OTT) was set up in July 2002 by DOE for the FreedomCAR and Vehicle Technologies Program.[26]

In addition to the FreedomCAR and Vehicle Technologies Program, tax privileges for consumers were also granted in an attempt to strengthen the EFA market. The federal government encouraged consumers to purchase EFAs by tax deduction.[27] A federal Internal Revenue Service's law of tax deduction arrangement for new purchase of personal, business, or trade used clean-fuel vehicles was offered to consumers from July 1, 1993 to December 31, 2003. It has since been extended to December 31, 2006. The deduction sum, sometimes as large as some $2,000, is subject to the incremental cost of clean-fuel burning equipment. New York State also offers such similar tax benefits for clean-fuel car procurement. However, the tax deduction has not facilitated the purchase of American EFAs, because, as mentioned above, American auto environmental technologies were not able to be commercialized according to the government-set schedule and were not able to meet the latest emission requirements either (not even the most advanced diesel technology can meet the demanding Californian regulation of NOx). Ironically, Toyota's Prius and other non-American clean-fuel cars have taken this tax deduction advantage. Owners of Honda Civic, Honda Insight, and Toyota Prius hybrid electric vehicles (HEVs) are qualified to claim the maximum deduction of $2,000 of the federal Clean Fuel Vehicle Tax Deduction. In effect, Japanese EFAT has "exclusively" enjoyed the tax breaks, whilst the tax deduction was open to each brand name and was primarily designed to encourage American consumers to purchase various new auto technologies such as hybrid, natural-gas, ethanol, and diesel, and hydrogen vehicles. For this reason (but the reason declared by the American authority was to avoid possible lawsuits arising from the water pollution of gasoline additive MTBE),[28] these tax breaks (as a part in the $31 billion energy bill) were terminated by Senate Republicans for the year 2003. Consumers who bought a hybrid car in 2004 would have their $2,000 tax deduction scaled down to $1,500 and in 2007 to none. For the replacement of these terminated tax breaks, a wide-coverage tax credit system of a similar amount ranging from battery electric vehicles, fuel cells, and alternative fuel buses, including alternative fuels and associated infrastructure costs was proposed by Bush's *Energy Plan* in 2003, attaching it to the Clean Efficient Autos Resulting from Advanced Car Technologies (CLEAR) Act.

6.5   Evolution and Achievement of the Japanese Environmentally Friendly Auto Technology Development

After the impact of the yen's sharp appreciation caused by the Plaza Accord in September 1985 (and by the fact that the U.S. dollar was ready to decline in any event), and the collapse of the bubble economy since the early 1990s, Japan has experienced a serious decline in car exports and a very sluggish domestic car market, forcing the world's second largest car production nation to restructure its industry to an extent never seen before in Japanese history. In addition, as Japan has to import 100 percent of its oil, of which about a quarter is used in the transport sector, the development of EFAT is very important. Accordingly, the government has regarded the fuel efficiency and low emission of autos as a top policy priority. Yet, the possible solutions seem diversified. Electric, LPG, CNG, hydrogen, and other alternative fuel vehicles have been recruited in the key programs of the government's energy policies. In this context, basic research for fuel cells, batteries, and other advanced automotive and transport systems are under way. However, the clear national goal and traditional Japanese industrial policy would not necessarily succeed in introducing EFAT to the consumer market. In the 1970s and 1980s, although Japan economically grew in leaps and bounds, the EFA market remained stagnant. MITI has been funding general fuel cell and other oil-alternative energy technology R&D programs for a long time,[29] planning and coordinating them through its subsidiary the New Energy and Industrial Technology Development Organization (NEDO) since the establishment of the latter in October 1980.[30] EFAT was however far from mature for commercialization over this decade.

In October 1990, the Japanese government through the Council of Ministers for Global Environment Conservation (CMGEC) introduced its "Action Program to Arrest Global Warming," emphasizing the reduction of green house gas emissions from transport vehicles for the period 1991 to 2010. The major Japanese automakers responded to this government plan also by announcing their own EFAT development plans in 1992, for example, Toyota's Earth Charter and Honda's Environmental Statement. Following this, the 1992 "Extraordinary Ordinance Regarding Reduction of Total Volume of Nitrogen Oxide Gas Emission by Automobiles in Specific Areas" and the 1993 "Guideline to Attain Restraint of Nitrogen Oxide Emissions from Automobiles in Specified Areas Relating to Transportation" were also prepared. Unlike the American government, the Japanese bureaucrats did not give the industry a deadline or push the automakers hard. Under this softer government attitude, Japanese automakers accelerated to search for higher but workable fuel efficiency targets. The government did not specify what kinds of EPA technologies should be prioritized. Rather, MITI provided around $250 million per year for technology that was to be quickly adopted by the market as the focus of government-industry cooperation to execute a similar activity and tailor-made technologies to fulfill the latest highly tightened requirement of the California market (also because the United States is the biggest market). Thus, the Japanese government placed emphasis on the reduction of nitrous oxide emissions (the crucial requirement of the California rules) in its government-industry R&D partnership programs. Apart from government sponsorship of the core emission free vehicle technology, other auxiliary technologies including advanced ICE, light automotive materials, manufacturing innovations, and electric hybrid vehicles were also included in the government budget. In the partnerships, the Japanese automakers that aimed at meeting the 1998 California ultra-low emissions requirement included Toyota, Honda, Mazda, and Nissan. Other Japanese automakers Daihatsu,

Mitsubishi, Isuzu, and Suzuki also joined forces with the Japanese government in R&D programs for different EFA technologies. Nevertheless, the Japanese auto industry has not developed a large and single cooperative R&D program emulating the USCAR and PNGV. In comparison, the Japanese public-private programs are dispersed and individually small in scale.

Having been developed by different automakers toward different technologies for the early 1990s, the foundations of Japanese EFAT were centered on hybrid technology, which associated modern petrol engines with electric motors to lower fuel consumption and emission levels without forfeiting power output, safety, and comfort. The launch of gas-electric hybrid vehicles such as the Toyota Prius or Honda Insight in the late 1990s and their mileages on roads have demonstrated the feasibility of the technology, paving the way for the full-scale introduction of zero-emission cars. The ambition and vision of gas-electric hybrid cars were contingent upon the possibility of a highly efficient battery to be recharged alongside with a gasoline tank. Thus, the government introduced further assorted fundamental partnership programs with different automakers. The public-private cooperation reinforced the significance of the hybrid engines based on their user-friendliness and high fuel efficiency. Such a hybrid system combined the battery with a conventional engine using a sophisticated control system to smooth out the switch-over of the two systems and reducing the overall weight and wind resistance to achieve high efficiency and performance. According to the first few Japanese gas-electric hybrid models, fuel consumption and CO_2 emission were about 50 percent lower than conventional, equivalent cars. The emission standard was far below even the lowest requirement of "Euro Four," which will take effect in 2005.[31] Gas-electric hybrid could use the existing refueling stations in any countries and it could automatically recharge the batteries, eliminating the previous problem of finding external electricity sources for recharge. The engine also automatically turned off when the car came to a standstill, so it never lavished fuel or created pollution while stranded in traffic. Moreover, the hybrid design was meant to change its primary fuel from gasoline over to fuel cells.

Toyota's Prius entered the domestic market in December 1997 and the export market in May 2000. The early version of Prius was comparable in size and specification to a Toyota Corolla, but about 30 percent more expensive in the purchase price. Nonetheless, the fuel reduction of almost 50 percent could help offset the purchase price. In June 2001, since the first day of sale with some 3,000 orders three-and-a-half years ago, the domestic sales of the Prius surpassed 50,000 and export sales amounted to nearly 300 vehicles to the state and city of New York. Up to early 2003, Toyota sold approximately 120,000 Prius hybrids and expanded its hybrid technology to other models such as the Estima and the Crown. Also, Toyota announced its intention to sell 100,000 units of the electrical unit of the hybrid's power train to Nissan to recover some of its development costs.[32] The latest version of Prius won 2004 Car of the Year award from *Motor Trend* magazine and "green" awards from many European countries. Toyota recently announced that its global cumulative sales of hybrid vehicles had exceeded 500,000 units by the end of October 2005.[33] Meanwhile, Honda was the first automaker to sell its hybrid Insight in the United States in December 1999, the unconventional Insight two-seater, and expanded the hybrid technology to its popular Civic in March 2002. While the Japanese EFAs met seven out of the ten requirements to achieve the best fuel efficiency specification listed by the U.S. Environmental Protection Agency, Honda's Insight was the first vehicle model, consuming only 60 mpg (one mpg better than the 59 mpg fuel rate of Toyota's Prius). Since 2003, Honda's Civic hybrid has become more popular than its other hybrid model Insight when the company has sold 10,607 Insight hybrids and 12,000 Civic hybrids in December 2002 since their launch.[34] Honda recently announced its global sales of hybrid vehicles exceeded 100,000 units as of April 2005.[35]

6.6   An Institutional Comparison between the United States and Japan

Using a broad institutional approach that highlights the difference between the United States and Japan, pitfalls of the American public-private partnership program are manifest. The institutional structure of PNGV brought about anti-incentive for and resistance to clean-fuel technology innovation. This institutional arrangement was composed by the following five layers:

Short-Term-Horizon versus Strategic Business

Since the American political institutions do not strategically coordinate product markets, the market-oriented automakers do not necessarily facilitate the PNGV's long-term, steady targets. The short-term, profit-making strategies of Detroit's Big Three made their goals gradually diverge from those of the federal government, even though their initial mutual interests prompted them to join the program working together.

With the general economic upsurge and the relatively low fuel price in the 1990s, consumers gradually preferred the more powerful SUVs for their daily use. Up to 1999, SUV sales rose to almost 19 percent of the total light vehicle market.[36] Due to the popularity of the expanding SUV market (especially those in the medium-size category), emphasizing power and utility rather than fuel efficiency, manufacturers did not want to lose this lucrative market segment and kept adding SUVs to their vehicle production line. The American automakers swiftly reacted to the profitable products because they were under pressure to maintain a high return on equity (ROE). Companies in Japan are mainly financed through loans in contrast to their U.S. competitors through private capital, leading to the phenomenon that stockholders do not demand for short-term profits. While Japanese investors could accept a lower ROE, for example, Toyota roving around 3.4 percent and 2.7 percent respectively in 1999 and 2003, in corporate yearly reports,[37] the American shareholders normally expected a much higher ROE, for example, GM maintaining 29.1 percent and 35.6 percent respectively in 1999 and 2003,[38] to reflect good corporate management. In addition, massive layoff of employees was the American way to improve corporate performance. For example, from 1991 to 1993, GM dismissed around 74,000 employees to maintain its profitability.[39] Likewise, despite a net profit of $570 million in the first half of 2002, Ford was prepared to close five plants and cut 21,000 jobs in North America to deal with slumping U.S. sales.[40] Under this commercial norm, the American automakers had no point to stake their resources in long-term, strategic investment, such as the EFAs in PNGV.

The American institutions also do not coordinate the process of cooperation and dialogue. The overall American automakers' independent corporate governance from one another and from government remained unchanged, notwithstanding participation in the public-private cooperative program for technological innovation. The three auto markers were not used to disclosing their on-going technological development, neither to one another nor to the public. A handful of reports were carried out by the independent NRC but they were focused only on the management of the program and compared different technologies in relation to the government's preset goals. They have not examined intensively the program design and process, let alone the technical information exchange for facilitating cooperation among the automakers. With the culture of commercial confidentiality on sensitive information and data especially the in-depth design of the new technology, and the cost of the technology investment, it is very difficult to carry out a real cooperation for information exchange, especially for innovation improvement. The traditional relationship among the three U.S. automotive companies Chrysler, Ford, and General Motors has long been one of competition rather than of partnership, adding to the challenge from several government agencies and national laboratories. In fact, the Big Three did not rely highly on the R&D funding offered by the U.S. government. For example, while the federal government had annually paid some $300 million in PNGV R&D fees, industry's total internal R&D expenditures (other than PNGV) amounted to about $12.2 billion annually. It implies that the government subsidies did not have a large impact on the Big Three's R&D budget. Don Walkowicz, executive director of USCAR, testified at the House on April 21, 1994 that "These additional costs borne entirely by industry include product and process design, certification for reliability and durability, safety and environmental testing, and investment in production machinery and tooling. These types of expenditures to productionize a successful development are typically on the order of 'four' times greater than the total R&D cost." Actually even the ratio related to the PNGV program amounted to forty times, that is, $12.2 billion versus $300 million.

Along with the neoliberal policy of the United States, American companies are used to streamlining their operations and productions. The car industry makes no exceptions. Most advanced (and risk-taking) innovation for the EFAT has been contracted out by American automakers. The relationship between the automakers and their sub-suppliers has been slowly transforming from vertical to horizontal form. That is, the manufacturing process from raw material to final product has been gradually faded out. The latest practice has been the outsourcing of more components, and automakers are more or less responsible only for assembly, sales and marketing, and distribution. Generally nowadays, from one-third to two-thirds of car components are purchased from suppliers. This contraction of investment commitment by contract-out has a preference for market-oriented and short-term innovation but would have impaired the long-term innovation investment strategy of the industry. By contrast, traditional Japanese automakers (except recently foreign-owned and foreign-operated ones such as Nissan) still comparatively keep a vertical relationship with their subsuppliers, even though some Japanese automakers such as Toyota have recently begun to acquire components from outside their *keiretsu* (groups of companies). Although the reliance on *keiretsu* suppliers has been reduced, especially after the economic bubble burst in the early 1990s, de facto exclusive or loyal suppliers are still often seen in the Japanese auto industry. For example, the relationship between Toyota and one of its most important suppliers, Denso, is not exclusive, but they have been working very closely with each other over the last few decades (Anderson 2003).

Incoherent versus United Parliament (or Congress)

A hostile political climate has hindered the progress of the American programs. As discussed in the last chapter, its political institutions provide opponents with legal and powerful tools to discourage the government from making EFT policies and from implementing relevant programs. The bipartite legislative structure does not promote a concerted political culture to support the American auto industry toward a common goal. Criticism, largely from the Republicans, had cast a suspicious eye over the partnership because it breached the usual principle of "small government." As soon as the bill for PNGV was approved in the House of Representatives, controversy followed immediately. The confrontational congressmen went through a time-consuming process in budgeting. The Republicans won the legislative vote over the Democrats to downsize the government's expenditure in November 1994. The PNGV budget was thus abated. The administration had to work hard to retain the PNGV budget but the congressional barriers had not yielded a facile allocation of PNGV funding in the following years. In the financial year of 1995, about $308 million was appropriated for ongoing PNGV-related R&D at eight federal agencies. Nevertheless, the legislative bodies always balked and refused to approve the administration's attempts to allocate governmental funding. For example, for the financial year of 1996, the administration demanded $383 million but the House assented to only $228 million, a reduction of $80 million from the $308 million appropriated in the financial year of 1995 and, even though the resolution of the Conference (meeting of two legislative committees) to settle the difference in the bill resulted in garnering $312 million, it was still $71 million lower than the amount the administration had requested. Following these hostile votes, the Republicans again acquired 214 votes over an opposing 211 votes in June 2000, winning in the House to largely reduce nearly half of all PNGV funding ($126.5 million), earmarking the PNGV funding as unnecessary subsidies to the wealthy automakers who earned nearly $20 billion in profits in 1999.[41] The House's Committee on the Budget and Congressmen such as John Kasich, John E. Sununu, and

Robert E. Andrews have publicly criticized that the PNGV funds were "corporate welfare."[42] Some even asserted that the PNGV program had been used by the automakers to avert tougher standards of CAFE,[43] based on the fact that CAFE standards would not be raised in the duration of the ten-year PNGV program.[44] The bipartisan parliament or congress kept arguing over this issue and continually disapproved of or slashed the PNGV budgets or bills, heavily impairing the research progress and investment confidence.

What is more, this contentious American congressional culture does not encourage long-term investment on politically risky products, unlike the more harmonious Japanese legislature dominated by Liberal Democratic Party (LDP).[45] Thus, it did not support the voluntary motives to search for technological excellence in the PNGV program. The erratic environmental policies in American political history remind the automakers that the goals of such a program might change or even terminate if the next election is won by the other political party. As a result of this anxiety, the automakers would not devote all their energy and resources to an area that a future government might not support. Although, this can be moderated by passing a long-term legislation such as the one that related to the semiconductor industry (one that supported a 10-year partnership), the auto making history (e.g., the refusal to obey the 1967 Air Quality Act and AAQS) tells us that this cannot be done if the government and the targeted have different goals and interests. Such a hostile political climate occurs because the car industry keeps trying to lobby their unwanted regulations. The industrial lobbying behavior suggests that the auto industry would not be bona fide in cooperating with the government in the PNGV program because industry was (and also is) trying hard to lobby Congress, aiming at increasing the CAFE standards and the zero-emission rules. Josephine Cooper, president of Alliance of Automotive Manufacturers, at a panel presentation of the 2000 Future Car Congress, Washington, DC, stated that "manufacturers do not like (it) when government takes the wheel in the form of mandates. . . . I never met a mandate I did like." In this generally hostile government-industry relationship on regulatory issues, the industry is not used to actively reducing emission levels or making fuel reduction improvement, so government officials should make the requirements more stringent in cases that incurred additional R&D and production costs.

In contrast, the Japanese policy regimes are more stable and coherent since the legislative assembly (Diet) has been dominated by LDP (usually in both houses since 1955). The stable and coherent policy regime is also attributed to the fact that Japan does not have a federal system like the one in the United States, and its 47 prefectures depend on the national government for subsidies. Once a regulation is adopted, it is fully supported by the majority of the Diet. The impact of such a political institutional structure on Japanese EFAT development is significant. Without this political support, the American political structure does not accommodate Japanese-style government-industry cooperation in which private investors have confidence giving them a favorable business environment for long-term ventures.

Oligopolistic versus Competing Auto Industry

The oligopolistic structure of PNGV could not motivate the automakers to innovative technology. This structure was an upshot of negotiation between the automakers' civil society and the Clinton-Gore administration. The joint-forces of the automakers were strong enough to protect their own benefits. As large but single public-private cooperation covered the holistic automaker, technical targets became identical to every automaker and the industry received government subsidies and research supports altogether. Thus, this quashed the need for further technical exploration beyond the targeted specification and did not provide the automakers with an atmosphere to compete against one another. This sort of oligopolistic collaboration even enabled the automakers to exempt themselves from antitrust laws and was inclined to slacken development of those risky and unprofitable products. The Japanese government, however, has not developed a large and single cooperative R&D program that emulates the USCAR and PNGV. In comparison, the Japanese public-private programs are dispersed and small. The amount of their budget each year was far smaller (equivalent to a total of about $35 million) than that of the American PNGV's $250 million because the Japanese strategy was only a kind of coordination, not subsidization (see budgets in table 6.2). In fact, the R&D expenditure of the Japanese auto industry supported by the government was not much. Usually the Japanese government/industry R&D spending ratio was about 1:4, while that of the American government/industry was about 3:1 so that the Japanese government spending on the R&D did not have a large financial impact on the auto industry (Polverini 1997: 5). For example, one of the most significant Japanese partnership projects, Advanced Clean Energy Vehicles (ACE), had a budget in 2001 of only 650 million yen, which was equivalent to around $5.4 million. Rather, as compared with the $260 million annual budget of the

PNGV program, the principal benefits of Japanese partnership were largely realized not via funding but primarily via government's careful and strategic coordination. The selection of what to be subsidized at a later stage betokens that the selected technology types are more important and have a potential to be accepted by regulators and consumers. Overall, the basic strategy of MITI's early policy in the EFA industry is, firstly, to support basic R&D to a moderate degree, highlighting those prospective technologies; secondly, to stimulate a competitive atmosphere; and thirdly, to induce the vitality of the industry through close cooperation with the industry with an active and flexible approach (Watanabe 2003). Through such an approach MITI induced a chain reaction that the automakers would accumulate their active and wide range of R&D (e.g., Honda's fuel cell vehicle design that uses the electric motor-drive portion of its electric cars, the fuel tank of its natural-gas cars, and the electricity-management technology from its hybrid cars).[46] The coordination of MITI helped the industry make use of the advantages of such incremental innovation. Watanabe further reported that "such an industrial policy system, in co-ordination with other related industrial policies, aims at inducing a chain reaction of the vitality of industry by stimulating industry's potential desire for active R&D. Technology complementation with capital as well as substitution for constrained production factors such as labour, energy and environmental capacity can thus proceed." Similarly, the participation of MITI through the above-mentioned *small* and *diffused* cooperative R&D programs with the auto industry was a fundamental effort, trying to elicit a lucid and ambitious technical commitment from the industry, unlike the way in which the American government fulfilled only the literal goals of the PNGV program.

Table 6.2 Major Programs and their Budgets in Japan

Name of Public-Private Cooperative R&D program	Fiscal Year Budget 1998 (million US$)	Fiscal Year Budget 1999 (million US$)
High Performance Batteries for Evs (1992–2001)	26.6	32.5
Polymer Electrolyte Fuel Cells (1992–2000)	3.6	9.5
Advanced Clean Energy Vehicles (ACE Project) (1997–2003)	4.1	6.0
ITS technology utilized for Clean Energy Vehicles (1999)	—	9.5

Source: IEAHEV.[47]

Slow versus Quick State Review

The PNGV program fell short in that it lacked a central, powerful coordination mechanism for the progress monitoring. The PNGV committee consisted of seven government agencies and the three big automakers, being monitored by OSG representatives from the U.S. Department of Commerce and the three automakers with a rotating chairman. A technical committee comprising of members from the three companies and seven government agencies was also formed and it reported to the PNGV committee, managing ten technical working groups of engineers and scientists. The general coordination and management were carried out by DOC and DOE (Chapman 1998: 18–25). However, without lawful and regulatory organizational frameworks, this sort of voluntary state-industry cooperation lacked the kind of bureaucratic leadership that enabled the management office to direct and motivate technological progress. On July 30, 1996, during the testimony before the House Science Committee on the PNGV program, Dr. Robert L. Hirsch, president of The Energy Technology Collaborative, Inc., indicated that "The program management office, located in the DOC, operates more as an information office than as a program directors office. The staff of the program management office has essentially no authority to impact the federal program other than through personal persuasion. (The) PNGV management at the DOE similarly is not empowered to directly influence the array of federal projects potentially contributing to the PNGV initiative." In fact, unlike the nonpolitically appointed and specialist-competent Japanese counterparts, the PNGV's government representatives did not have the authority to effectively allocate the funding to the highest specification technologies. The U.S. technical officials might have expertise to discern the viability or applicability of the proposed technical solution from their industrial partners, but they were legally powerless to make decision or to influence the industry's choices. Certainly, the adjustment power of DOC or DOE was far lower than that of the bureaucracy of Japan. Meanwhile, an independent third party was appointed to regularly review and comment on the management and technical progress of the PNGV program: the National Research Council (NRC), an affiliate of the U.S. National Academy of Sciences and the National Academy of Engineering.[48] The NRC had reviewed the PNGV program only once a year since 1994.

In contrast, in Japan, bureaucracies are entitled to rule over the economy and most industries by exercising the powers of controlling foreign exchange and investment, and of creating and administrating cartels, import barriers, and export subsidies. With these legal powers, Japanese government officials are able to give regular and influential administrative guidance (*gyosei shido*)—without the need for debates in parliament—on important industrial issues such as price, production, or export levels. Although not all of these traditional bureaucratic powers are still applicable or useful to regulate the automakers over the last decade, the bureaucrats have played an important role in making sure that the EFA developmental process of the government-industry programs are in line with national goals.

On the American side, since the formation of the PNGV program and its technological goals in 1993, the three automakers were given freedom to select technologies to fulfill those goals. The PNGV committee in the United States did not have a formal mechanism such as *gyosei shido* to correct the intermediate goals and trajectory if they went astray or were found necessary to adjust in order to resolve mid-course challenges. Instead, being managed by those powerless American bureaucrats, the responsibility to provide unbiased, constructive suggestions for improvement to the PNGV program was on the shoulders of the NRC, which also had neither economic power nor legal authority but merely evaluated progress and suggested solutions to problems once a year. In fact, *gyosei shido* is still one of the most powerful capabilities of government to motivate industries in Japan; although one might argue that the advantages of pluralistic evaluation in the Diet would be missing, the case shows that, without this, the recommendation for improvement of the PNGV program had to be made through time-consuming testimonials and political debates in Congress. The DOC officials could not take immediate administrative power as it was what the Japanese bureaucrats usually did to quickly rectify any off-course developments, leading to a very inflexible developmental process. As a result, the PNGV goals became stifled and focused only on family-style car size with equivalent performance and lower emissions—thus missing the chance to develop hybrid electric designs or fuel-cell engines that might better suit the latest more stringent requirements of zero-emission mandates. For example, the regulation fulfillment pressure drove the PNGV committee to shelve the previous research findings of battery electric vehicles, findings that were conducted by the U.S. Advanced Battery Consortia (USABC, a product of former public-private cooperative research between the three Detroit's automakers and the electric utility industry). The NRC committee reported that "If the USABC had viewed the EV not only as a competitor with the gasoline-fuelled ICE (internal combustion engine) vehicles (but also the possibility of being a complement), it might have established more attainable performance goals."[49] Indeed, the government officials of the USABC were program goal administrators and not catalysts for changing or combining the two technologies, because goal modification had to be made by legislators. Without quick, incessant examination of the programs by bureaucracy, the PNGV goals were regulation-like and not beneficial for rapid, incremental technological development.

Powerless versus Discerning Bureaucrats

The program review and adjustment system through legislators was also questionable in the United States. The only official review was through NRC reports on the PNGV research program by the end of each year (the first report was issued in October 1994). In addition, the PNGV program itself lacked authoritatively powerful administrators to achieve a deliberative and quick adjustment system. For this, the PNGV program did not have formal public-private linkages at all. The progress-monitoring and improvement system for the PNGV partnership was located in Congress rather than in government agencies and did not constitute a frequent and extensive consultation mechanism for reviewing and adjusting programs and regulation by bureaucrats.

In contrast, institutionalized public-private dialogues through advisory councils (*shingikai*) and brain-trust politics (*burēn seiji*) have been introduced since 1970 when the first LDP president, Nakasone, wanted to obtain public support for counteracting the great influence of other factions inside the LDP. "Japan boasts the most extensive set of institutional arrangements for reaching agreement between government and industry. The bureaucracy can count on more than 200 deliberating councils-forums for public-private consultation on key policy issues" (Weiss 1998: 55). In fact, instead of congressional inquires, dialogues between bureaucrats and automakers that are followed by administrative orders to adjust the course of policy were often seen in Japan and were the theme at the early EPA technology development stage. In response to a directive from MITI, in contrast to the inflexible, coercive laws from the American Congress, requesting Japanese automakers to produce HEVs to be scheduled for production in 2000, on January 16, 1992 Toyota Motor Corporation (TMC), in its Earth Charter document, initiated a corporate target for the development of lowest-emission vehicles, similar to that of the American PNGV program. A research committee, G21, was formed under the leadership of Corporate Chairman Eiji Toyoda in September 1993. Similarly, Honda set up a Honda Environmental Committee (HEC) in 1991 and in 1992 announced its Honda Environment Statement (a statement that would comprehensively address Honda's environmental plans and restructure its organization toward meeting its goals).

Nevertheless, the Japanese bureaucrats through their close contact with the automakers found that it was not the time for the government to heavily subsidize manufacturers or consumers at this stage, even though they have seen that their American counterparts spent millions of dollars helping the automakers. Rather, in addition to the basic R&D research as mentioned above, governmental agencies have regularly and closely cooperated with G21, HEC, and other industrial specialists and associations in information exchange and standardization activities. For example, the New Energy and Industrial Technology Development Organization (NEDO), since its formation in October 1980 as a governmental agency under MITI, has served the auto industry through its world information data base, library, and publishing activities. The NEDO Information Center is accountable for providing up-to-date world information through its official connection with the framework of the International Energy Agency (IEA), in addition to the market information sharing among their *keiretsu*. Nevertheless, the information provision efforts are only to support the growth of the industry, not to make them share individual technologies with one another.

Two years later at the 1995 Tokyo Motor Show, the G21 team exhibited the first concept hybrid car model Prius. In June 1995, the exclusive technology of the hybrid powertrain of Prius was named "Toyota Hybrid System" (THS). In May 1997, the first media test-drive demonstration was carried out and, later in October of the same year, Prius was officially revealed to the press at the Tokyo Auto Show. Two months later, Prius was the first low-emission vehicle (LEV) available for sale in Japan's auto market, gaining a record 18,000 sales in the first business year, followed closely by the commercial launch of Honda's Insight hybrid. To support these fresh technologies, the bureaucracies found a need to intensify its publicity. After discussing with the industry, the government extensively organized introduction and promotion activities for them. For example, in 1999, Toyota cooperated with the government by participating in a total of 65 exhibitions, including the Low-Pollution Vehicle Fair held by the Japan Environment Agency, and the Clean Energy Vehicle Popularization Project Test-Drive Meeting held by NEDO and regional bureaus and departments of MITI. Likewise, in 2000, Honda cooperated with national and local governments in some 24 exhibitions and lectures on EFAs in all parts of Japan. Meanwhile, the Japanese Agency of Natural Resources and Energy (NRE) of MITI was developing unified standards for fuel and its infrastructures, working hard to establish a legal and regulatory framework for the use of alternative fuels gradually shifting to emission-free fuel cells.[50] In December 1999, NRE actively organized a deliberation panel consisting of representatives from industries, academia, and government, aiming at unifying standards for fuel cells and their corresponding items. Overall, the Japanese public-private cooperative EFA programs have been supported and adjusted swiftly by bureaucracies who are empowered to judge and determine, and not by legislators (as is the case with the PNGV in the United States).

The establishment of different emission goals at different stages needed regular public-private communication to make manifest a realistic and progressive technological process. Although the American automakers might find that there was more room to go beyond the goals (compliance responsibility was already fulfilled), they did not have the motivation to exceed the goals and thus retarded their efforts or reduced further investment in emission-level reduction programs. Such retarded performance could be detected by an effective, frequent public-private connectedness and refined quickly by virtuous bureaucratic power. For example, as diesel hybrid system was already able to meet the scheduled 1997 "downselect" requirements, there was no point for the automakers to decide on other more costly, though less polluting, designs such as fuel cell, compact hydrogen storage, ultra-capacitors, or even gasoline-electric hybrids. When the worst came to the worst, the powerless American bureaucrats were unable to redress quickly the "deflections" that were not in line with the latest emission regulations elsewhere. As a result, the diesel-electric hybrids were unfortunately unable to meet the national tier two and Super-Ultra-Low-Emission Vehicle (SULEV) standards of California,[51] even though the specifications had met the PNGV requirements. The exhaust gas of these diesel-electric hybrids was far dirtier than the nondiesel designs and was not a solution of oil shortage problem either.

Despite the fact that DOC and DOE were in charge of the program and the overall goals were conceived by governmental officials, technical developments were left to industry (Chapman 1998: 17). So were the Japanese counterparts. However, the government-led overall specifications were not always so realistic (these were only predictions of possible technology in ten years' time) and regular and frequent deliberations had to be conducted to select appropriate *intermediate* procedures/technologies for fulfilling or refining the goals of the national policy. In the presence of powerless American bureaucrats, despite the participation of governmental officials, the partnership structure permitted the industry to decide on the interim technology selections without lawfully rectifiable bureaucratic surveillance, resulting in a choice of technologies inevitably in favor of the profit-making industry. For example, the decision on compression ignition direct injection and the diesel-electric hybrids was shortsighted because these technologies only mechanically served the purpose of the emission goals but were not as a long-term solution. Therefore, the industry would have just regarded the government as a regulator but not as genuine cooperative partner for technological advancement. While the industry had such an attitude solely to meet preset standards, an innovative spirit would not be composed. Questions such as why 80, and not 100, miles per gallon; why high fuel efficiency, not zero-emission level; why 10 years, not a shorter period would not be raised and investigated by the industry. And these were exactly the immediate tasks that the American government officials should have assumed, but they did not.

In contrast, the coordination by the Japanese government is a step-by-step approach and very selective in judging. During the early stage, the government refrained itself from emphasizing any EFAT. Rather, it conducted a series of public-private cooperative R&D programs working with the auto industry, ranging from high performance batteries, polymer electrolyte fuel cells, and advanced clean energy vehicles to intelligent transport system (ITS) technology (see programs in table 6.2). Although the former two R&D programs (for the lithium secondary battery for EV and the fuel cells) had started since 1992, neither electric vehicle nor fuel cell vehicle was chosen by Toyota or Honda for sale in the market in 1997. The other two early R&D public-private cooperative programs were for advance energy vehicle technologies, of which engines and engine storage systems for six prototype vehicles using alternative fuels such as methanol, natural gas, and dimethyl ether commenced in 1997, and ITS technology started in 1999. Thus, these two "late" programs are not a factor of the "early" success of Toyota's Prius and Honda's Insight. This suggests that the gas-electric hybrid system was developed by the automakers themselves (not by the early decision of the government as is the case in the United States), while such a hybrid system was not included in the early government-industry cooperative programs. Only at this very moment, the Japanese government found a need to quickly adjust its efforts to support the hybrid technologies.

Entering into a second phase, as soon as the gas-electric hybrid models emerged in 1997, the Japanese government initiated subsidies for consumers to spur up the market (table 6.3). The NEDO program granted car buyers this subsidy, paying the difference between the costs of a clean energy car and a conventional gas or diesel car with equivalent power output and performance. The early subsidy was confined to those who used the vehicle on the job. From the second year (1999), the subsidy extended to those who commuted to their workplaces. This subsidy resulted in a quick expansion of the hybrid sales since its inception in December 1997 (table 6.4).

Furthermore, when the government found there were considerably mature and feasible technologies available in the market, it believed it was the right time to offer consumers a greater deal of incentive and promotion for the purchase of the new technologies. Since 2001, national incentives for LEVs under Income Tax and Corporation followed by various local taxes were offered. In addition, also from 2001, a long list of subsidies for LEVs was announced (table 6.5). In parallel, the Japanese government could steeply increase emissions-control laws.

An individualized incubation effort of the Japanese government gradually appeared. To avoid duplicated efforts and rent seeking, the

Table 6.3   Government Subsidies for Consumers

Name of Subsidy	Fiscal Year Budget 1998 (million US$)	Fiscal Year Budget 1999 (million US$)
Subsidies for Consumers to Purchase Clean Energy Cars (1998–2000)	74.7	95.5

Source: IEAHEV.

Table 6.4   Hybrid Sales Increase Spurred by Government Subsidies

Name of Subsidy	Number of Vehicles 1998	Number of Vehicles 1999
Gas-Electric Hybrid	3,967	4,950

Source: IEAHEV.

Table 6.5   Subsidies for Low-Emission Vehicles in 2001

Governmental Institutions	Subsidies
Ministry of Environment	Subsidies for Priority Introduction of LEVs alternated with HD Diesel Vehicles
Ministry of Environment	Subsidy for Purchasing Air Environmental Patrol Cars
Ministry of Environment	LEV Diffusion Plan and LEV Diffusion Support Plan
Ministry of Economy Trade and Industry	Clean Energy Vehicles Diffusion Plan
Ministry of Economy Trade and Industry	Regional New Energy Introduction Promotion Plan
Ministry of Land Infrastructure and	Pioneer LEVs Evaluation Plan Transport
Ministry of Land Infrastructure and	Low-emission Trucks Introduction Promotion Plan Transport

Source: IEAHEV.

Japanese bureaucracy did not heavily subsidize the whole auto industry but only coordinated a clear division of development area. The ACE project has offered separate assignments to each automaker. For example, Nissan has been assigned series HEV methanol fuel cell for small vans; Nissan Diesel series HEV engine and capacitor for buses; Honda series HEV ANG engine and flywheel battery for passenger cars; Isuzu series HEV engine and capacitor for trucks; Mitsubishi combines HEV engine and Li-ion battery for trucks; Hino combined HEV engine and supercapacitor for buses. Even though the goal of these projects was to achieve better fuel use, each project has its own detailed target and timeline. On the whole, the Japanese public-private cooperation is merely between the government and individual automaker. The Japanese bureaucracy is responsible for ensuring that each participating company has its own focus. As soon as the Japanese bureaucrats ascertained where the strengths of the Japanese automakers were, more public assistance would be rendered accordingly. In contrast, American PNGV had allowed each company to carry out research on all technologies hand in hand. The American bureaucrats are not required to distinguish the leaders from the laggards for further incubation. Unlike the American PNGV program collectively heading for the 80 mpg target, cooperation between Japanese automakers was seldom because their public-private partnership programs had their own targets and research information to make technological progress independently. In addition, the government also encourages this kind of intra and indirect contention by emphasizing national goals as a means of fostering innovation.

The Japanese and United States institutional layers are thus summarized and shown in charts 6.1 and 6.2 respectively.

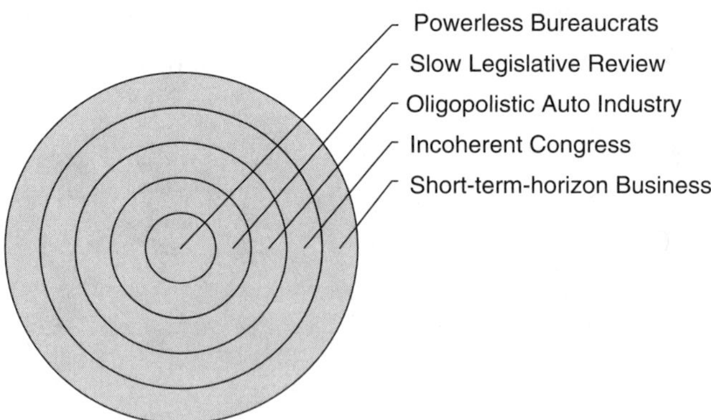

Chart 6.1   Five Institutional Layers of the American EFA Programs.

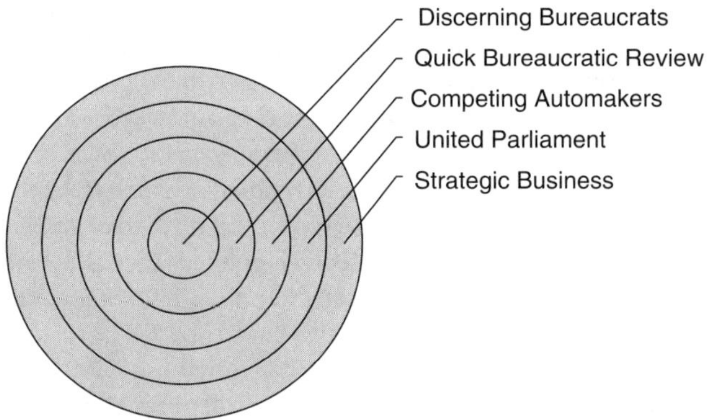

Chart 6.2   Five Institutional Layers of the Japanese EFA Programs.

6.7   CONCLUSIONS

This chapter has examined the "black box" of public-private partnerships by comparing the government-industry cooperation in the
United States and Japan. The two countries appeared similar in
automotive public-private cooperation, though they pursued different courses. The American public-private partnership was restricted
by the formation of inflexible program goals. The automakers, who
treated the inflexible goals as costly regulations, tended to keep the
goals unchanged, leading to inelastic technological levels of development deviating from a proper track of progress. An inefficient
review through yearly reports and congressional debates and the
horizontal and slack public-private linkages account for the cessation
of the program.

Japan's EFA public-private partnership programs since the early
1990s have emerged as industrial responses to the CMGEC's 10-year
program. These include "The Action Program to Arrest Global
Warming" and other onward government environmental laws and
regulations (e.g., the 1993 "Guideline to Attain Restraint of Nitrogen
Oxide Emissions from Automobiles in Specified Areas Relating to
Transportation") that were treated as an early signal of a new industrial
transformation. In this context, the fast growth of EFATs was not really
achieved by government physical efforts and subsidies but by the
automakers' own endeavors and resources. The Japanese government-
industry partnerships acted like a foundation and gave clear guidelines
and a congenial environment for the automakers to further develop
technologies. However, not every Japanese automaker is able to effectively

manage their own resources for advancing their environmental technologies. For example, while Nissan, Mazda, and Mitsubishi were still traditionally confined themselves to those expensive and inflexible products from their *keiretsu*, Toyota and Honda were sourcing more parts from outside of their *keiretsu*. In their capacity as monitors and arbiters, the bureaucrats were able to decide which technologies/automakers should be greatly supported in the next cooperative phase, so as to boost popularity of their selected technologies. The increasing government efforts to bolster the selected automakers and the subsidies to consumers who bought the hybrid models have shown the subtle Japanese government-industry relationship. In addition, the Japanese automakers were more willing to develop the costly EFAT than their American counterparts were. This is because the Japanese government did not worsen conflicts of interest between conventional auto business and EFA technologies by moves as refraining from pursuing a fixed target, reducing the risk of EFA business through dispersed and small public-private programs, and financially supporting some consumers to buy the new products. Then, government efforts were concentrated on technologies or companies that had the highest potential. This answers why Japanese automakers are more receptive to government EFA policies; and why Toyota and Honda were the only two leading Japanese EFA producers in the early stage of the EFA development; and why their technologies are more advanced than their U.S. counterparts.

The facts imply that a government-industry joint-entity or a high degree of state power is not necessarily effective for an industrial sector to produce synergistic outcomes. The success of such public-private cooperative program is dependent upon its overall institutional arrangement, that is, the structure of public-private program itself and the national institutional environment it interacts with. In addition, state power has to be gradually and selectively employed, especially in dealing with conflicts between different goals and interests of government and industry. The American and Japanese institutional infrastructures can be generally classified into two types: liberal market and coordinated market. That is, the American public-private cooperative programs were managed by market-oriented corporate governance and dichotomous congressional debates, whereas the Japanese public-private partnership programs were supported by bureaucratic coordination and in a coherent and stable political system. As long as state capacities are established, it is suggested that the Japanese style of public-private cooperation is more effective than the American counterparts. An argument has been put forward that effective public-private cooperation for EFAT development rests more on state capacities than on market capacities. In turn, according to the facts in the comparative case, state capacities depend upon efficient, quality, and concerted bureaucratic autonomy rather than upon time-consuming, political, and argumentative parliamentary or congressional legislature. Quick, powerful, and remedial actions to rectify partnership program errors, mistakes, or abuses that are harmful to national benefits are essential. Yet, in the EFA sector, having to smooth over the conflict between corporate interest and regulatory pressure, an effective public-private partnership is achieved by a softened and step-by-step, rather than coercive and holistic, governance attitude, and by following the aforesaid state capacity concepts.

From the questions posed in this chapter, the answers to this comparative study are clear. Coordinated market institutions offer considerable advantages, but only in countries such as Japan where the latest tweaked institutions have quality state capacities, particularly in terms of balanced, tailor-made, and increasing support to automakers (and not to all members of the target industry regardless of their individual capability). This not only answers the queries of Sperling (2000) but also complements the findings of Wallace (1995), Evans (1997), and Weiss (1998) on public-private cooperation. In today's public-private confrontational environmental economy, the step-by-step, flexible state capacities (better achieved by "the rule of bureaucracy" than by "the rule of law") in this Japanese case are far more effective than market capacities and can facilitate a quick diffusion of environmentally friendly technologies.

An editor in *The New York Times*, October 1993, commenting on the PNGV program wrote that "Corporate statism arrived in America last week in the back seat of a dream car. . . . Ironically, we're aping Japan just when the terrible weakness of the protective Japanese system is becoming manifest."[52] However, the Japanese system in the case study did not prove to be protective any more, it was rather developmental since its institutions were just tweaked and optimized. Public-private cooperation has long been the "trademark" of developmental states, such as Japan, Taiwan, and South Korea, during the postwar period. Since the economic surge in these countries after World War II and the recent publications of studies and reports (e.g., the World Bank's World Development Report in 1997) on the developmental role of governmental institutions, the neoclassical countries have begun to apply their energy to examining the virtue of state involvement and capacity (even though the 1997 Asian Financial Crisis and the prolonged recession in Japan have revealed the "dark side" of the developmental state and cooled down the heat). EFAT policy is one of the examples where newly revised state involvements that do not compromise the advantage of market forces have been seen in many successful policy sectors.

7

Conclusions

Summary Statement of Objectives

The purpose of this book has been to analyze and account for the significant environmental-technology (E-T) developments that have remarkably emerged since the early 1980s. A systematic comparison of varying institutional structures across advanced industrialized countries and time periods was the means by which this book's study was conducted. Developments in the wind energy industry in the United Kingdom and Germany, in the packaging waste recycling industry across the United States, the United Kingdom, Australia, Germany, Sweden, and Switzerland, and in the environmentally friendly auto industry in the United States and Japan were compared and presented in chapters 4, 5, and 6 respectively. This chapter includes both the discussion of the results and findings.

The theories of state capacity led us to infer that although countries have similar E-T policy goals and program features in general, they have different policy approaches, patterns, and outcomes due to different state involvements and institutions. This inference has further deduced three propositions related to the impact of state institutional capacity on early E-T developments. First, E-T diffusion can proceed from neither market-oriented "deregulation," nor industry-led "self-regulation," nor "coercive regulation," but rather from dynamic "obliging regulation." The adjective "obliging" is used to describe the soothing effects to the "anger and pain" of polluting industries who are unwilling to transform themselves. Second, while the governments of the advanced industrial countries have employed "obliging" state power and adopted certain degrees of state involvement to achieve E-T diffusion, they performed in different ways due to their different state-societal institutional infrastructures (i.e., state and business structures, and their relationship). Third, "pacifying permeation," which is a subtle form of state-societal relationship signifying a dual concept of "pacifying" and "permeating" state efforts, has driven the E-T development and has successfully permeated the polluter-dominated industry with E-Ts by alleviating sufferers' resource and financial burden. In other words, government's "pacifying permeation" efforts can create a close association and indirect interdependent relation between the polluter-dominated industry and E-T developer when one (e.g., wind developer or recycler or greener automaker) receives benefits (e.g., subsidies or favorable, guaranteed payments or purchase orders) directly or indirectly from the other (e.g., electric utility or packaging industry or polluting automaker) without damaging the latter's business.

To start with the comparative case studies, I enquired into the infrastructure of institutions, centering the institutional difference between the sampled countries. According to the characteristics of the E-T industry, I compared their different degrees and extents of regulations, state-societal structures and relationships, and investigated what sorts of institutions are more effective for E-T development. Secondly, I analyzed whether there are common features of the E-T leaders, trying to explore the sources of their policy capacity. Generalizations were extracted from the data of case studies for answering the research questions and implications were drawn for establishing new assertions. On the whole, the comparative case studies gave rise to new insights about the role of bureaucrats and E-T business developers. The case studies attempted to analyze three E-T policy sectors in different countries at different times and in different places, featuring and contrasting their common regularities and irregularities that tested the new state capacity concepts and also developed new ideas of state governance.

Overall Results and Findings

To sum up the outcomes of the three comparative case studies, the principal argument is strengthened. That is, effective E-T development rests upon state capacities, as essential complementary elements to market capacities, that can better develop technologies or economies, which are prone to conflicts of interest, and, thus, increase institutional comparative advantages to change industrial behavior. In turn, the three propositions are also supported by the outcomes.

It is confirmed that regulations to coordinate and fix corporate behavior, competition mechanism, risk and information sharing, market and systemic failures are essential, but a more strategic approach is required to deal with the increasingly powerful and defiant industry. More specifically, state capacities are composed of and depend upon an efficient and discerning bureaucratic autonomy rather than upon a time-consuming, political, and argumentative parliamentary or congressional legislature. Furthermore, quick, powerful, and remedial actions to rectify a deviated trajectory from national benefits through intimate public-private communication are also essential. In the cases studied in this volume, these state capacities were produced by a specific, dynamic institutional arrangement that allowed government officials to participate in tailor-making, implementing, and rectifying rules and regulations for specific actors. To explain this by means of a simile, introducing E-Ts to an unruly industry by a nation-state is like navigating ships in a turbulent harbor by a pilot. Therefore, E-T development is always better "navigated" by a "pilot" who has the acumen and shrewdness to make use of its own institutional advantages. In the three case studies, the position of such a "pilot" was successfully assumed by the bureaucracy using "administrative guidance" without a need to consult with their legislature, particularly as far as new, fast, innovative technological development was concerned. For example, it is found in the Japanese case that MITI has played a very important role in the EFAT development. Similarly, the European governments have also heavily taken advantage of their coalition government structure and their executive power at the prelegislation stage. Even the British and Australian governments in time learned to use more executive power by putting forward devolved or secondary legislation (the British Parallel Statutory Instruments and the Australian Regulations by ministerial decree) under powers given to them by international treaties or other foreign affairs.

Yet, these state capacities built by means of bureaucratic power do not mean that bureaucrats should assume all duties and responsibilities of E-T programs. Instead, such bureaucratic power is merely to deploy a workable regulatory framework and to decide if it is necessary to share the work load with the industries or how to support them according to the strength of each party.

The case studies further discovered that bureaucratic power was better used to monitor and control the progress of the shared duties. Particularly, having to smooth over the conflict between corporate interest and regulatory pressure, bureaucratic power can be used to choose the most potential and cooperative actors to work with in the first place. Nevertheless, this maneuverability of bureaucracy was built to the extent that nobody would seriously suffer from the laws and regulations. For example, the German government officials decided not to mechanically excise the law but were willing to compromise with the utilities despite winning the wind energy lawsuits; the

European governments were willing to make concessions, postpone mandatory deposits, and even lower the residual waste amount limit and recycling fee when they found the recycling requirements were too "harmful"; the Japanese bureaucrats did not put forward unrealistic, detailed targets from the outset of the EFA programs and refrained from creating losers by tailor-making individual, small public-private partnerships.

Furthermore, the results of my three comparative case studies show that, in the leading E-T countries, the relationship between government and polluter-dominated industry is one of interdependence, as is the relationship between government and the E-T developer. But the relationship between polluter-dominated industry and the E-T developer is *not* of interdependence. The term "pacifying permeation" is thus used to describe a relationship between E-T developer and polluter-dominated industry in which the former derives benefits from the association while the latter remains unharmed and unaffected. This relationship is deliberately created by governments in E-T leading countries. So this situation is neither zero-sum nor win-win, but a relationship that is somewhere in between. The E-T premium is largely absorbed by governments and consumers rather than only by industries in the E-T leading sectors. Illustrations of this phenomenon are abundantly available in the cases. Some of them are drawn out below:

- The wind power case shows that the German government has successfully motivated the tempestuous utilities to adopt wind energy by amending the EFL in April 1998, asking each level of the utilities to equally share the additional cost by limiting the financial charges of the different utilities to 5 percent each; and by allowing the latter in turn to charge residential consumers for the additional costs.
- The packaging waste recycling (PWR) case shows that the Swiss government has relieved the industry's container deposit legislation (CDL) financial burden by various government arrangements. The industry was not the only one to shoulder the cost of recycling, households too had to pay a "ground fee" for waste recycling service and "bin-liner fee" for municipal waste management and collection. Empty bottles are collected with the help of communal authorities. Thus, the cost can be further shared by local governments and households.
- The environmentally friendly auto technology (EFAT) case shows that the Japanese government relieved the auto industry from resource and financial pressure by refraining from putting forward

unrealistic, inflexible technical targets from the outset of the EFA programs, reducing the risk of EFA business through dispersed, small, and tailor-made partnership programs, and financially supporting some consumers to buy the new products, etc.

Finally, these governments have closely manipulated their regulations and even gradually reduced E-T premiums on industries. For a clear example, in Switzerland, owing to fast growth of collection volume and a better efficiency of handling, the recycling fee was reduced to 5 Rappen in 1996 and was further reduced to 4 Rappen in 2000. Thus, government efforts that have successfully relieved industries from resource and financial burdens so as to enable them to willingly adopt E-Ts have supported the idea of "pacifying" sufferers. Also, the word "permeation" underscores the fact that effective E-T policies are those that are applied only to cooperative market players first and then spread through them to other actors. This can be proved by Germany's new strategy, after facing formidable opposition from electric utilities, to concentrate on the most willing, small rural communities first and then to gradually expand to the utilities. The European coalition governments successfully regulated their beverage bottle industries also by using such a "permeation" method. They did it taking short steps and expanded the scope carefully. Sometimes they even retreated a (small) bit and watched, lest the regulations should become "unrealistic." They permitted industries of a single packaging tupe monopolize the collection of their own containers and did not allow them to extend the PWER systems to other packaging types until the pilot industry-run programs had made considerable progress. For example, Sweden developed aluminum PWR in 1984 and then expanded to PET in 1991, and glass in 1996. During the course of the "permeation," the governments made various attentive and supportive efforts to ease the industrial pressure of both resource and finance. By contrast, E-T policies in the E-T laggard countries, particularly in the United States, are applied to uncooperative actors, being nonpermeable. Thus, in addition to the word "pacifying," the successful policy tactic in building state-industry relationship can be called "pacifying permeation."

In sum, it is unequivocal that successful E-T developing governments concentrate their incubation effort upon the most accessible actors, while "pacifying" the most suffering actors. The intensity of state power increases little by little, ensuring a stimulated rate of progress with relatively low levels of conflict. In addition, the political and business institutions have a clear division of labor, arranging the best performance for the desired outcomes. The responsible bureaucrats are regularly monitoring the course of development if it derails from the right track. Government officials will exert quick and proper actions to remedy any mistake or deviation. A tailor-made program for each firm or actor group may avoid interfirm conflicts but create an intrafirm competition effect, stimulating actors to have a positive attitude and motivating them to continuously pursue technological innovation and diffusion. After supported actors make progress and unsupported actors change attitude, government effort can shift or enlarge to bring about permeation.

To analyze the composition of an effective institutional arrangement, I initiated a series of comparisons between LMEs and CMEs over time, in which I discovered that the regulatory framework is central to E-T development but has been changing in response to the need of each development stage. Also, the institutional structures and relationships among government, industries, and E-T actors accelerate or impede E-T innovation and diffusion at different times. There is a clear contrast in effectiveness between policies based on the two different economic models at the early development stage. The institutional configuration of LMEs was effecting higher levels of stimulation under generous government subsidization but was much less conducive to E-T development than was the case in CMEs when the subsidization was removed. However, E-T development in CME countries was sometimes impeded by the institutional network not adjusting to new challenges. It implies that only proper institutional arrangements at the right time can be an accelerator for E-T development. It also implies that there is no "one" best or fixed type of institutional arrangement for E-T development for all time. For example, although the regulatory frameworks of the United Kingdom in the early 2000s and Germany in the late 1990s are distinctively different institutions, both of them have effectively motivated their own industries to adopt wind energy. Policy capacity employed for E-T developmental purposes can be advanced only by compatible institutions at different stages. Therefore, an efficacious institutional arrangement is a multistep approach. This arrangement features a regular public-private interaction and a changeable institutional positioning. These two features do not connote different strategies but a dynamic institutional approach for E-T development, that is, an institutional arrangement that is subject to increase (and sometimes decrease) state power and the target area according to the actual situation in order to permeate the market.

A *dynamic* public-private relationship is a flexible way of producing progressive joint-force between government, industries, and other social groups. It presents a clear allocation of work, based on a *moderate* state power refraining from hostile action. The relationship is dynamic because the intensity of state support/power gradually increases (or decreases) as soon as progress is made (or hindered). Governments are gradually changing their attitude from encouragement to enforcement, achieving an effect of technological innovation and diffusion. As such, the form of state power sometimes can be a threat to the policy actors, but sometimes in a lenient way, which is in distinct contrast to the idea of "disciplined support" described by Weiss (1988: 73–75). The notion of flexible and dynamic state power of "pacifying permeation" can refer to a paradoxical policy strategy, that is, it is sometimes aggressive but at other times regressive. This strategy that the state be able to exercise considerable "maneuverability" (guiding business by a skilled movement and a controlled change of course) is quite as important as the strategy that the state be able to effectively deal with the problems produced by powerful, nonnegotiable, or rent-seeking policy actors.

On the whole, the generalized policy strategy can be simply described as tweaking institutions in the shadow of the state. A summary of the evidence is thus constructed in table 7.1.

COMPARISON OF FINDINGS WITH PREVIOUS RESEARCH

My results compare favorably with Vogel's arguments (1996: 3–4) that, first, "re-regulation," not "deregulation," is largely employed by the governments of advanced industrial countries to carry out reform and achieve their targets; second, different degrees and types of regulation were adopted by different countries; third, the reform locomotive is driven by states, not private interest groups. In turn, going by the comparisons of evidence made in this book, it is found that although at the early stage in the 1980s "deregulation" was firstly employed among the Anglo-Saxon countries, governments reapplied regulations quickly as soon as the outcomes of "deregulation" in the E-T area were found to discourage development. Meanwhile, although the sampled European countries and Japan had regulated their E-T markets at an early stage, they did not make progress until their early regulations were adapted to become more effective in solving the real problems. Furthermore, governments are without doubt the drivers of the reform in these cases. In this regard, the results of the case studies have perfectly matched the arguments of Vogel.

Table 7.1   Summary of the Evidence

	Evidence		
Propositions	Wind Energy	Packaging Waste Recycling	Environmentally Friendly Auto Technology
E-T diffusion can proceed from neither market-oriented "deregulation," nor industry-led "self-regulation," nor "coercive regulation," but rather from dynamic "obliging regulation."	The British regulation-free NFFO/SRO in the early 1990s resulted in serious contract implementation problems and local opposition but the resumption of helpful regulations and arrangements since the early 2000s through the Merchant Wind Power Initiative, the Climate Change Levy, the tradable Renewable Obligations, etc., and the new linkage between the government and local communities improved its wind energy diffusion by a sharp increase. The delegation of power to the German industry through the "indirect-specific program" in the early 1990s enabled utilities to refuse expensive wind energy, but the introduction of many technical, social and financial friendly regulations such as the fixed selling price of wind energy have boosted its wind diffusion to become the world leader since the late 1990s.	Repeal efforts in the United States have been very popular as the inflexible, coercive PWR regulations were always averse to the benefits of the industry. More flexible, considerate regulations have improved PWR rates in the other five countries. For examples, The 1999 amendments of lower PWR rate requirements to the British Producer Responsibility Obligations (Packaging Waste); the Australian National Packaging Covenant replacing hostile regulations; German industries to be exempted from their individual obligation to take back sales packaging by joining DSD; regulations sharing the Swedish industry's collection burden by requesting consumers to return their empty cans back to the RVM; regulations relieving Swiss industries from financial pressure by asking households to pay a "ground fee" and "bin-liner fee."	The U.S. confrontational regulations in the 1960s had led to a series of litigations between the federal government and the auto industry. Similarly, the American public-private partnership programs in the 1990s were restricted by the formation of inflexible goals. The automakers, who treated the inflexible goals as costly regulations, tended to keep the goals unchanged, leading to an inelastic technological development deviating from a proper track of progress. The Japanese regulations were refrained from instructing the industry what the best technology should be. Instead, the government waited and observed which technologies and/or automakers should be greatly supported in the next cooperative phase and then helped the short-listed auto models and/or actors by supportive regulations.

While employing dynamic "obliging regulation" to achieve E-T diffusion, governments did them in different ways due to their different state-societal structures.	Based on competition and other market-oriented policies, the British government has consecutively resorted to tendering systems, independent private producers, and tradable green certificates to establish a competitive, entrepreneurial institutional structure. As a result, the new regulations and market instruments facilitated the development of large firms and projects.	The powerful, argumentative Anglo-Saxon legislature at different levels hindered the governments (worst in the United States, moderate in Australia and fair in the United Kingdom) from enacting higher PWR requirements.	The American equity-dominated corporate structure with high ROI requirement drove automakers to concentrate upon profitable car models rather than unpopular EFA technologies.
	Based on market coordination, cooperation, and standard-setting, the German government has employed specific technical, social and financial regulations and arrangements such as a wind energy conversion system and a fixed selling price to the grid, providing a lucrative business for the rural communities. As a result, the friendly business environment attracted a large group of small, dependent, cooperative wind energy developers in terms of interconnected, affable local cooperatives or individuals.	The compromising, unitary European legislatures facilitated the European governments to produce concerted PWR enactments.	The time-consuming, argumentative and dichotomous parties in the American Congress without being supported by a powerful bureaucracy could not immediately identify the need for new program goals until a new government was formed.
		The majoritarian Anglo-Saxon government structure gave rise to repeals and oppositions from nongovernment legislators.	The Japanese debt-based corporate structure with low ROI requirement allowed automakers to pay more time and energy on the EFA technologies.
		The coalition European government structure (much in Switzerland, more in Sweden and less in Germany) enabled smooth PWR policy making.	The efficient, quality Japanese bureaucracy supported by its LDP- dominated parliament could achieve quick, powerful and remedial actions to rectify any deviated trajectory of program from national goals and interests.
	The British government invited small tenders for early wind projects. Its trading scheme has relieved unwilling actors from obligation since 2002 by allowing them to buy green certificate, i.e., concentrating on willing actors first.	State-societal dialogues via British ACP since 2002, Australian NPC Council 1999, German DSD 1991, Swedish *Returpack* 1984, and Swiss *IGORA* 1989 have improved PWR rates	The American government did not differentiate the automakers/technologies to carry out the PNGV partnership program, failing to incubate other marketable technologies.
		The British government improved the PWR development by being more considerate toward the industry, e.g., the 1999 amendments to the Producer Responsibility Obligations lowering some PWR rate, and adjusting PWR yearly.	

Continued

Table 7.1 Continued

Propositions	Evidence		
	Wind Energy	Packaging Waste Recycling	Environmentally Friendly Auto Technology
	In addition, the government resolved the local opposition by inviting local participation in regional strategies and targets for wind projects.	Besides, the government targets encouraged willing actors to adopt more recycling and relieved unwilling ones from the Obligation by allowing them to sell or buy PRNs.	Without government efforts to reduce automakers' opportunity cost of investing in such niche market and strong opposition force, the industry regarded the PNGV program as a costly burden to bear.
A subtle form of state-societal relationship has successfully permeated the polluter-dominated industry with E-Ts by push-starting the most potential actors while comforting sufferers.	Meanwhile, the British government prudently increased national target yearly from 3per cent in 2002/03 to 4.3 per cent in 2003/04 and so on. The German government amended the EFL allowing the most suffering utilities to equally share the premium with other levels of utilities and allowed them in turn to charge residential consumers for the additional costs. Meanwhile, the revision of German laws and regulations put more emphasizes on small and dispersed local communities who had less opportunity cost of producing more costly wind power. The government established an attractive, reliable investment environment for these local communities and spread wind energy diffusion throughout them.	In addition to efforts to relieve the industry's pressure. Germany chose PET in 1989 as the first item to be recycled. Then different quotas were given to industries. Although some finally could not meet their quotas in 1995, a long grace period until 2003 was given. The Swedish government developed aluminum PWR in 1984 and then expanded to PET in 1991 and glass in 1996 with considerate government supports. The Swiss government first developed aluminum in 1989 and then PET in 1990. Although industries finally did not meet its PET targets, the government did not penalize them but even lowered the residual waste amount limit and reduced the recycling fee to five Rappen in 1996 and further down to four Rappen in 2000.	The Japanese government relieved the auto industry from resource and financial pressure by refraining from pursuing fixed, unrealistic technical targets, reducing the risk of EFA business through dispersed and small development programs, and financially supporting some consumers to buy the new products, etc. As soon as the most potential technologies, i.e., gas-electric hybrid, and companies, i.e., Toyota and Honda outstandingly appeared, the government efforts were enhanced and concentrated on them. The Japanese rule of state power in terms of powerful and efficient bureaucracy and supportive Diet was more flexible and quick to review and improve the progress and to avoid from groundlessly regulating the auto industry, successfully permeating the technologies into the society.

However, the concept of re-regulation in Vogel's literature is far too general to describe how successful E-T development can be achieved. For example, whether new regulations employed by government are imposed on the industry/society or delegated by government to be self-regulated by the industry/society may lead to very different outcomes. Thus, this book has attempted to explain with further evidence how regulation has been employed to introduce reforms. This has also explored further dimensions of the regulatory state literature (Moran 2001a&b/2002; Cook, Kirkpatrick, Minogue, and Parker eds. 2004; Jordana and Levi-Faur eds. 2004/2005). Apparently, the concept of "obliging" regulation shares the view of a more light-handed regulation and steering role of the state. Yet, the governments of the leading E-T countries did behave submissively with the strong industries/societies. The formidable "bottom-up" forces of defiance were somehow detoured away by the states and only the cooperative "bottom-up" forces remained behind.

Taking this point into consideration, I may also compare my case studies with the data found in the literature written by Wallace (1995: 4) who argued that the revisionist view believing in the peaceful achievement of social progress through reforms by regulation was useful but too general to be understood by most E-T policymakers.[1] The case studies in this book have shown that the settlement of conflicts of interest in policy sectors such as E-T development cannot be effected by aggressive methods such as the government lawsuits against the American auto industry in the 1960s or the coercive laws and regulations that faced serious opposition and repeated repeals in the American PWR case. Further, concentrating only upon the kind of regulatory instrument that should be engaged can only help produce robust quantitative solutions regarding R&D or intellectual property rights, but, it has not much to do with E-T diffusion if state or social structures allow E-T opponents to fight back or shield them from E-T adoption. Wallace's case studies of different national environmental policies on industries have revealed that a positive, quality relationship between government and industry is beneficial to the industry's environmental reform; and by exploring cases among Denmark, France, Germany, Japan, the Netherlands, and the United States, he adds that environmental regulations in some countries have been a source of conflict but not in other countries (1995: 238). In this regard, my comparative case studies also indicate the same finding. He further suggested that a stable environmental policy and quality dialogue between industry and regulator are key themes and concluded that "the political, institutional and even personal relationship between policymakers and industry have the greatest bearing on (environmental technology) innovation" (1995: 5).

My case studies are in complete accordance with the arguments and findings of Wallace. Nevertheless, Wallace's case studies were too brief, especially the composition of the public-private relationship, because he had not investigated how governments dealt with the most serious tensions between corporate and national interests. A positive and quality public-private relationship wherein both government and industry mutually aim at the same E-T goal is not always attainable in every country, not even in closely government-industry-linked Japan. For example, in 1949 Toyota defied the government and restructured its own auto business; in 1963, Honda expanded its business to the auto-making industry without paying heed to government disapproval; in 1969 Mitsubishi defied the government and entered into a venture with Chrysler.[2] In addition, even if on occasion good public-industry relations exist, the heavy opportunity costs of transformation and/or short-term profit-oriented corporate management may still hinder polluting industries from achieving the envisaged national E-T development goals. These problems have to be resolved by government's special tactics. Therefore, a closer investigation as to how government can effectively direct policies onto the right track is necessary. Wallace's research was done in the early 1990s when outstanding progress of E-T development was yet to be seen. My comparisons between and observations on the earlier sluggish and recently fast E-T diffusion in the three E-T sectors have provided more detailed context to unravel these puzzles.

A few scholars have recently argued the case for regulatory (and institutional) innovation in order to bring about more efficient diffusion of new renewable energy technologies (e.g., Isoard and Soria 2001; Johnson and Jacobsson 2000/2002; Bergek 2002; Madlener and Jochem 2002). In view of the increasing importance of learning effects and economies of scale and scope, future new E-T projects will need to be developed by scaling down, that is, by avoiding starting with too large a project. The case studies in this book, especially the one dealing with EFA technology, share these findings. Significantly, Japan's policy decision to encourage the simultaneous development of a variety of small-scale EFA programs gives credence to the conclusion of Johnson and Jacobsson (2002: 31) that "technological uncertainty is high and industry needs to place its bets widely in terms of experimenting with a variety of designs." Despite this, some of these scholars have indicated that "[many actors have) seen the diffusion of renewable energy technologies as a threat" (Isoard and Soria 2001: 631

in Madlener and Jochem 2002: 5), and that "the long-term transition to sustainable energy systems needs . . . the simultaneous activation of latent drivers and a reduction/elimination of existing barriers" (Madlener and Jochem 2002: 6). My thesis shares these views and shows extensively that, in terms of the E-T leading countries it samples, the solution may well lie in the application of specific laws and regulations, and state-societal structures and relationships.

Although Wade's (1990) main argument was the importance of state power and the causal relationship between governed market policies and the outstanding industrial performance in his three East Asian cases, he concluded that "the three countries (Japan, South Korea, and Taiwan) established institutional arrangements to avoid the pitfall of government officials taking the lead role in a new industry while knowing little about it. Government officials are involved in a policy network with sources of information much closer to the operational level of particular industries" (1990: 336). This is confirmed by my findings that successful E-T developments were monitored, coordinated, and adjusted by bureaucrats who intensively communicated with target industries at operational levels. Further reference is especially made to the Japanese case of automobiles in the literature from Wade (1990), in which, as an example, he denied the argument of Schultze (1983) who claimed MITI's intervention was a mistake since Japanese state power failed to consolidate the domestic automakers in the 1960s. Wade defended his advocacy of state intervention by pointing out the findings of Taizo (1984) that MITI's active interventions and efforts in the auto components industry in the 1950s and 1960s were helping the unconsolidated assemblers to increase their global competitiveness. Wade (1990: 330) further disagreed with Schultze's statement by restating the findings of Thurow (1985: 244) that "it [Schultze's argument] ignores the fact that the 'mistake' never occurred, because MITI backed off in face of the firms' resistance; and it ignores the effect of the 'mistake' on the unconsolidated firms' investment and strategy, which was to redouble their efforts to prove MITI wrong and themselves right because they knew that if they were not successful MITI would be back with a new consolidation plan at a later date." In other words, this has confirmed that the early Japanese auto policy strategy conformed to the initially "pushy" attitude of the governments studied in this book.

However, the aim of this book is to discover not only the optimal stance that governments should adopt toward E-T development but also in how governments should deal with the opponents to adopting E-T policy. This book has found that not every industry was afraid of a "pushy" government. The results in the three case studies demonstrate this point further since most early E-T policies did not benefit the industry and a GM strategy faced strong resistance and jeopardized technology innovation and diffusion (e.g., German wind energy market before the mid-1990s; American packaging waste recycling market over the last four decades). Therefore, Wade's GM theory is only part of the story (i.e., the early stage of developing countries) but cannot fit to other conditions, especially the recent mature economic development requiring a higher degree of public-private cooperation. The emphasis on GM theory by Wade to explain his three cases was obviously influenced by his minute attention to the historical background of Japan, Korea, and Taiwan, looking back even to the tradition and aftermath of samurai bureaucracies in the Tokugawa period and various regional and world wars, concluding that "Hard states are able not only to resist private demands but actively to shape the economy and society" (1990: 337–342). This book does not deny that a hard state was one of the most effective policies especially in north-eastern Asia during their chaotic periods of rebuilding. But, when industries have matured and become increasingly organized, capitalized, and powerful, a hard state force would inevitably face strong resistance, for example, quite a few Japanese industries such as consumer electronics and auto production, since they became more organized and economically powerful than ever before, have from time to time defied the bureaucrats' administrative guidance. Wade (1990: 336–337) nevertheless admitted that "However, we know rather little about how the information that lies with the parts of the bureaucracy in intimate contact with particular industries is flushed up to the level of strategic, government-wide decision-making." My case studies have considerably filled this gap.

Furthermore, my data also supports the results of Evans's case studies (1995: 247–249) showing that robust, unconstrained bureaucracies connected to social groups and classes are essential for national economic development, that is, his concept of "embedded autonomy" (EA). My cases have clearly shown that governments have to work closely with industries/communities. In the PWR cases, after discussing with the private sectors how a recycling program should be done, the European governments delegated most of the work (i.e., waste collection system design and operation) to the civil societies while still participating in some complementary jobs (e.g., additional collection points and container parks to ensure an easily accessible infrastructure for consumers to return their consumed bottles). The governments then kept a close eye on progress and forced the industries to pay a mandatory deposit if the industry did not fulfill the recycling goals in time. These cases show that the key for success was not only intimate public-private cooperation but the sober, vigilant, and persistent attitude of the government officials toward higher recycling rates. This confirms that a balance between embeddedness and autonomy exists in successful packaging waste recycling adopters. In contrast, the arm's-length, public-private relationship in the American recycling case has demonstrated the scenario described by Evans that the neoliberal and (neo-)utilitarian visions of both governments and consumer goods producers have made for a very weak state "embeddedness" (1995: 22–25).[3]

Meanwhile, strong political lobbying in the U.S. recycling case has also led to very weak state "autonomy." Thus, the packaging recycling law- and regulation-making was frequently abolished by clientelistic political pressures and by the plurality of industrial interests. Although the American specially established PNGV program was a government-industry partnership, the government still failed to galvanize an unwilling and powerful industry into developing E-Ts. It is also because the partnership of the PNGV was carried out by the American bureaucrats who were not endowed with sufficient power to review and modify the program under the powerful American legislative structure. Hence, an inefficient and incoherent institutional setting can be a disadvantageous factor in the functioning of "embedded autonomy." For example, the American EFA technology development that relied on unilateral policymaking, a third party to review progress, and an argumentative parliament to adjust the deviation caused the pace of its EFA technology development to severely lag behind (contrary to their other vehicle technologies being decided and developed by only automakers themselves). Rather, the Japanese bureaucratic apparatus that was equipped with sufficient authoritative power to carry out the making, implementation, and correction of policies without being held back by the political system was able to achieve a much better performance. The comparison between the U.S. and Japanese EFAT cases suggests that "embedded autonomy" can be better achieved if both market- and bureaucracy-compatible institutions are established.

Weiss (1998) has contested the idea that "globalization" is inimical to state capacity for domestic economic transformation. Instead, Weiss's comparative analyses of various countries have indicated that strong state transformative capability has played an important role in the process of "globalization" through strongly constructed domestic public-private linkages. Weiss found the possession of a robust government-industry relationship described as "governed interdependence" (GI) to be of much advantage. My studies on the state response to regional/global pressures on E-T innovation and diffusion have also found the importance of maintaining a specific government-industry relationship. The question as to whether a state can successfully transform its industries depends upon its transformative capability. My cases additionally show that, however, generally industries were very reluctant to adopt E-Ts, not to mention to strike a deal or to arrange a quality cooperative mechanism with the government. Even though particular industries were willing to form partnership with government, it did not automatically result in a satisfactory result. This situation not only occurs in the liberal market economy (LME) cases but also in some coordinated market economy (CME) countries. For example, German electric utilities have been fighting off expensive wind technology by suing the government. The overall picture indicates that previous forms of government-industry interdependence did not exist in the E-T industry with fast growing business. Among the four forms of GI, that is, disciplined support, public risk absorption, private-sector governance, and public-private alliances, none was evident in the successful cases of this book. The four forms of GI conceptualized by Weiss were successful governance modes in many industries but were not effective development tactics in E-T sectors. Therefore, "pacifying permeation," a new form of state-societal relationship or policy network theory (or an additional form of GI theory), is better used to describe the state-societal relationship in E-T leading sectors.

Another outstanding feature of my cases is that all the successful E-T developing countries have more effectively executed their programs and more smoothly achieved their goals through *dynamic* public-private relationships, not *static* public-private partnerships or joint-ventures. In fact, the successful development process was not achieved by a constant public-private partnership, but by different types of relationships at different stages. Even though an intimate government-industry relationship is highly important all the time, government is not required to become a stakeholder or a semientrepreneurial participant. The cardinal quality of a synergetic public-private relationship is defined not only by its initial organizational and target settings but also by its capability of adjustment (of process vis-à-vis the market change). The flexibility and adjustability of targets or quotas or programs or government roles in the two successful European wind energy cases (the United Kingdom since the early 2000s and Germany since the mid-1990s), the three successful European PWR cases and the recent fast catching-up British and Australian PWR rates, and the successful Japanese EFA technology case have all been seen and they have fitted in with the conclusion of OECD's literature on Local Partnerships for Better Governance that

"Evidently, it is not because a network of partnership is established that policy coordination automatically improves. *Adjustments* must be made to maximize the effect of partnerships (relationship) on governance" (2001a: 126). On the one hand, its focus of adjustability and flexibility can be achieved by "the needs of local public service offices" for a more flexible management of programs, ensuring better "involvement in its implementation" as suggested at the last point in its conclusion (2001a: 127). This "hardware" for a better flexibility of management is useful because it is observed that the governments in the successful E-T cases have built different local facilities at different stages in order to assist project process, stay abreast of latest progress, and solve problems locally and directly, even though they were not stake holders in projects. On the other hand, "software," that is, "bureaucratic maneuverability," for a better flexibility of governance is also needed and found in my cases. It means that, the role of government or public-private relationship changes at different stages so that E-T programs can be effectively and efficiently adjusted in due course (e.g., the Japanese EFA case). In E-T leading sectors, where bureaucrats are embedded, autonomous, powerful, and knowledgeable, a high quality "bureaucratic maneuverability" can be found.

This finding of public-private partnership is also shared by Wolfgang Polt's *The Role of Governments in Networking* in the OECD's literature (2001b). He summed up the network-oriented policies for Austria, Denmark, Finland, Greece, Hungary, Italy, Switzerland, and Taiwan carried out by the OECD Focus Group on Innovative Firms and Networks and generalized some tentative conclusions. Polt (OECD 2001b: 313–315) has found not only that "many countries have identified collaboration, cooperation and networking-oriented policies as important technology and innovation policy instruments" but also that "Policies need to be *tailored* to the different needs, incentives and capabilities of the participants." These findings have properly explained the recent trend toward more efficiency-enhancing government intervention and public-private partnerships in the innovation-deficient policy sectors. My case studies have also contained these facts (e.g., the tailor-made public-private cooperative programs in the EFA technology sector in Japan). The American PNGV, which was only a general but not a tailor-made partnership for the strengths of each automaker, had a less satisfactory performance than those Japanese individual programs that supported each automaker separately and specifically. Polt at the end of his conclusion shared his findings with those of Guellec and van Pottelsberghe (2000) that "policies have to provide a *stable* environment in terms of funding and institutional setting." Truly, a stable, supportive, and considerate government role in giving tailor-made incentive and assistance in due course is an intervening variable also for E-T development.

Having assimilated the concept of stable political environment but dynamic state power, the idea of a quality public-private relationship has become particular rather than general. The discovery of the importance of the particular public-private relationship in this study has implications for understanding the failure of some public-private relationships. Callon (1995) found that MITI's finely promoted industrial policy was not necessarily successful all the time. The MITI-sponsored research and development aimed at assisting Japan's information technology industry to reach the cutting edge level failed in the 1980s due to interfirm conflict. The EFA technology cases have confirmed Callon's finding by showing that, the horizontal (interfirm) and fixed type of the American public-private partnership has repeated MITI's previous mistake by neglecting interfirm conflicts (to share innovations with one another), while the vertical (intrafirm) and tailor-made Japanese public-private partnership achieved fast technological progress by avoiding reluctant interfirm sharing. Therefore, a successful form of a particular public-private relationship possesses tailor-made programs that are monitored and adjusted by bureaucrats to develop the company's strengths.

Conclusions based on the Findings

Over the last two decades, E-T has been developed under starkly different economic models. Free market theorists argue that government can advance the national interest through policies that emphasize "liberalization" and "deregulation." According to these two principles of free market economics, more competition within a market should be introduced and government regulations should be eliminated. Meanwhile, coordinated market theorists assert that government can succeed in developing the technology through cooperative government-business relations. They also argue that state-led market power supersedes neoliberal market forces and that domestic institutions are critical to economic development.

This book adopts an institutional approach to examine whether some political economies have greater impact on a country's environmental technology development than others by systematically comparing and analyzing the different policy approaches and outcomes between liberal and coordinated market economies. Wind power, PWR, and EFA technology were chosen for the comparative case studies because they are the most mature, influential environmental technologies. The data for the three case studies were collected from the United States, the United Kingdom, Australia, Germany, Sweden, Switzerland, and Japan, because they have remarkably different institutional structures and outcomes.

Results indicated that in the early stage, the development problems, patterns, and outcomes of the Anglo-Saxon economies were very different from the others. These distinctions can be explained in terms of institutional effects. Nevertheless, at a later stage, many of these countries have improved their unsatisfactory outcomes by modifying their own institutions with more "obliging" regulations and "discerning" bureaucratic power. I thus maintain that domestic institutions and the bureaucracy play a very important role in environmental technology development. As the policy patterns and outcomes do not vary considerably from one country to the other, a generalized explanation can be formulated.

EXPLANATION OF THE FINDINGS

The findings first suggest that regulated institutions are fundamental to E-T development. Deregulated institutions through the implementation of market-oriented competition programs, through the delegation of powers to industry, or even through the encouragement of public-private partnerships do not necessarily lead to the growth of E-Ts. Regulation relying on state power motivates E-T activities and initiates progress but creates the *mise-en-scène* for public-private confrontations; though deregulation relying on market enjoys a boom in terms of player numbers, its undesirable secondary effects drive E-Ts into decline. Nevertheless, at a later stage, many of the sampled countries have learnt the lessons from their earlier experiences. Deregulation for competition policy and regulation for compliance with E-T program standards have not only become important but have also been shown to coexist productively. But each sampled country has its own combination of deregulation and regulation. The new finding is no longer focusing on whether the Anglo-Saxon countries should use more regulations to eliminate E-T development problems to which market forces do not have a solution, or whether the European countries and Japan should delegate more power to or cooperate with their industries to utilize expertise and experience that government regulations alone cannot provide. Instead, the role of governments in a particular institutional structure, and the combination of deregulation/regulation must be strategically applied/adjusted according to the competitive advantages of the institutions at a given stage and time, leading to a common policy strategy: "pacifying permeation."

The primary policy technique is to know how to select potential E-Ts and cultivate them much like experienced gardeners who know that vine plants must be pruned if they want to produce better quality and more abundant fruit. Thus, the role of governments in the successful E-T cases can be compared to a grape grower; it is simply unwise to grow grapes on unfertile ground. At first a government must "sow seeds" in a particular area and then nurture through fair amounts of subsidies and/or privileged treatment. Unlike grapevines that climb on tendrils and have to be pruned hard for maximum yield, juice quality, and good vine balance, opposition from the industry and other social groups should be "pacified," not "pruned," that is to say E-T development has to be supported by political institutions and government officials have to be trained to be able to identify actors who defy the regulations and/or abuse the subsidization schemes and support. And in doing so, government will able to select the most cooperative actors and provide them with greater strength, quality, and balance. All poorly developed E-Ts except the "strong" ones are later removed from support. The "strong" E-Ts are guided upward as the main runners. When selecting E-Ts, the bureaucracy usually "picks" those that are pointing in the direction government wants the E-Ts to grow. This resembles bureaucratic power, rather than legislative power in rectifying mistakes quickly without being prolonged and politicized. The resulting E-Ts develop strong sustainability taking after a mature grapevine to bear fruit and to build without support. As soon as this successful governance and development process is readily accepted by others, diffusion of the technology will take place.

Implications for E-T Policymaking

Policymakers in Western countries used to enjoy the advantages of "the rule of law" envied by other nations, especially developing countries and even Japan. After a centuries-long practice of making laws by parliamentary legislators, it seems that the era of courageous, powerful government is well and truly over. Bureaucratic hands are tied and their feet are bound due to their traditionally powerful legislatures. This fact is, unfortunately, neglected by many people and thus policymakers and state institutions are profoundly misunderstood and distrusted by citizens in most countries.

The world has recently begun to suffer from the severe challenges of environmental degradation and resource scarcity. People expect a mature parliamentary system to be able to address every problem in a country and to obtain the best solution by carrying out thorough discussions in a representative democracy. People are also afraid of the possible abuse of bureaucratic power and wrong decision-making. Yet, the past decade of environmental-technology lag in countries governed by "the rule of law" has left many of us more disappointed than ever. However, the recent efforts to catch-up by British wind energy and PWR developments and the Australian PWR industries after managing to rely on more bureaucratic power than parliamentary power, as shown in this book, indicate that their market-oriented economic model bears little relationship to the failure of their previous E-T development efforts. People who believe in the supremacy of CMEs over than LMEs (or the other way round) for E-T development are barking up the wrong tree. When E-T policymaking is based on the influence of voters, tabloids, and interest groups, the issue of national E-T development would become a political game. "On the day the British Opposition leader announced he would rein in the expansion of wind farms if elected, the (Australian) Treasurer, Peter Costello, was in South Gippsland, assuring locals that 'wind turbines are ugly and it should be up to (local) councils to determine whether or not they are allowed' " (Van Tiggelen 2004). Due to the restriction of the present "rules of the game," that is, the overreliance on parliamentary systems, politicians have no choice but to react for survival. This explains why politicians in these countries continue to dither without a long-term, persistent commitment toward E-T development. In addition, the response of legislatures to problems is far slower than that of bureaucracy. Thus, "the rule of law" is unable to deal with issues of high-tech developments that require rapid responses from the state in the increasingly competitive market of today. Therefore, a reform of state structures to achieve a better balance between "the rule of law" and "the rule of bureaucracy" is required for the imminent era of environmental economics.

IMPLICATIONS FOR FURTHER WORK

The contributions in this book have provided a combination of concepts of institutional arrangements and strategies to resolve the initial E-T developmental puzzle. The case studies have demonstrated that state capacities are essential to kick-start an early environmental economy. Further studies on similar policy logjams should be conducted to further analyze the new argument and concept in this book and to examine whether they are also applicable to other industries (of the same and other countries) having similar characteristics on E-T development.

Additionally I recommend further comparison between "market capacities" and "state capacities" for the next phases of E-T development. Although this book has shown that successful E-T development in the initial stages must rely on dynamic state capacities, further studies on ongoing E-T developments are equally essential as the process of E-T development is not yet complete. More mature E-T actors such as the publicly listed Danish windmill producers will require further investigation to show whether "market capacities" are needed when E-Ts become considerably mature and how the selection or combination of "market capacities" and "state capacities" can help conventional economies transform themselves into sustainable environmental economies, without subsidization.

To this extent, the evidence over the last few decades has shown that both LME and CME institutions have invested a great deal of effort and vast sums of money into E-T development, especially those that involve strong state involvement. The findings thus also question the effectiveness of such vast sums of money in many ongoing E-T subsidization schemes. While some critics may say that the large diffusion of the E-Ts in some countries is the result of the heavy supply-side policy of governments, others are of the opposite view. As Freeman Ford, who set up FAFCO, America's oldest and biggest producer of solar pool-heating equipment, commented, "In the early '80s, the market got up to about $600 million in sales—that was several hundred thousand domestic hot-water heaters that turned out to be a tax-credit bubble." According to Reuters (January 25, 2002), Danish Economy Minister Bendt Bentsen commented that "We are very concerned about the costs for society and for Denmark's competitiveness if we continue to expand the use of green energy." In any case, the original idea of these E-T developing regimes asserted that once the E-Ts became mature enough, government subsidies that were provided to stimulate initial demands could be removed. This book has proved this assertion by finding that some of the subsidies in the case studies have been really removed and that the subsidies were not the locomotive drive of the E-T development. For example, the Danish wind energy investment subsidy of 30 percent (excluding the large utilities), granted since 1979, was terminated in 1989 when the demand became strong enough to survive without it. Today, the German and Danish wind energy markets and the Japanese EFA markets are more reliant on government's supportive rules and arrangements than on directive financial assistances and incentives (these E-T leading countries spent much less on subsidization to the market than those laggards). Therefore, this general debate should be further researched, considering measurement of financial, environmental, and/or social gains instead of E-T quantitative uptakes as a basis for comparison across different institutional structures. Such a research is worth conducting and can be treated as confirmation of the significance of institutions and state capacities. Of course, they have to be carried out on a much larger scale so that a fair and accurate measurement of financial, environmental, and/or social gains may be obtained. As most of the E-Ts over the last few decades have only been in their infancy, further study on what has happened in the mid- and even late 2000s will be equally informative and stimulating.

Furthermore the findings in the EFA case study imply that the latest Japanese mechanisms of state governance may not necessarily be suitable only for catch-up economies. This is contrary to other findings in the literature on Japan's economic model. There are quite a number of scholars describing the Japanese economic model as "catch-up capitalism," adding that the model of Japanese governance is limited to its decades-long commitment to catching up with other advanced industrial countries by means of innovation through imitation. For example, Yamamura (2003) argues that despite intensive reforms since the economic bubble burst in the early 1990s, Japan's economy still remains catch-up industrialization and developmentalism. Thus, this book can be seen also as an effort to open research agendas in multiple high-tech fields that are governed by more bureaucratic power (through tweaking institutions) supplementing the insufficiency of market forces.

Appendices: Definition of Terms

Appendix I   Types of Environmental Technologies

E-Ts have evolved from the very basic filtering devices, catalytic converters, desulphurization, and NOx removal technologies to the environmental-friendly equipment and techniques such as low-emission engines (lean-burning) or low-emission fuel technologies (liquefied petroleum/natural gas or compressed natural gas). The concept of E-Ts has also been modified from environment-improving knowledge and devices to environmental applications for the industry, particularly global, commercial, nonwasteful, and permanent applications. Renewable and recycling technologies are seemingly the targets of many present scientific and political front-runners. Of course, there are still many other scientific dimensions and areas to search and make further E-T breakthroughs. But, if they are still under scientific research or are not yet mature enough to be launched in the market (e.g., the environmental use of nanotechnologies or superdurable materials), I exclude them from my case studies. My study focuses on the business-oriented type with a commercial market track record of at least a decade, this is the basis of all my study across different countries and their institutions. I define and categorize the types of E-T as follows.

Biosphere-Conservation-Oriented Environmental Technologies

Technology is defined as "systematic application of knowledge to practical tasks in industry" in the Oxford Dictionary. Alternatively, Green and Morphet (1977) define technology as the systematic knowledge of technique, which is the interaction of person(s)/tool or machine/object that defines a way of doing a particular task. Generally, new technologies are devised to increase the productivity of various production factors. However, traditional E-Ts are related to protecting ecosystems by preventing pollution (and controlling pollution that is not entirely preventable) and by remedying contamination that has already happened. To conserve the environment or to maintain the ecosystems of the earth, it is important to rediscover the balance of nature and to reintegrate the diversity of technological traditions by different forms and means of E-Ts.

Thus, while E-Ts have obvious material applications, the greater importance of E-Ts lies in the principles and methods, or the know-how generated by such technologies. For example, this includes the knowledge to reduce pollutants and to recycle them, the production processes to regenerate and renew resources, and the know-how to replace the nonenvironmental products by environmental ones without generating waste. From this, focusing on biosphere conservation without considering commercial sustainability, the primary E-Ts are defined as "the practical application of knowledge determined by science to solve environmental problems."

Business-Oriented Environmental Technologies

Nevertheless, the concept of E-Ts has to be changed, inasmuch as the mantra of sustainability is being chanted nowadays. That is to say, the new target of E-Ts is to be energy-efficient, globally competitive, value-added, or productivity-enhancing rather than merely environment-protecting. According to the logic of globalization, E-Ts are designed to increase global competitiveness by arranging resources to maximize efficiency. Technologies provide techniques to reduce the amount of usage of natural resources that become more expensive than ever because of environmental degradation and energy shortage. Technologies allow corporations to produce products by ecological means that greatly reduce production costs. From this point of view, any practical application of environmental knowledge or production method or managerial technique that is integrated into, transforms, or replaces the industries (and possibly enhance productivity), or increases competitiveness, adds value, or creates demand is regarded as a technology for contemporary environmental needs: business-oriented E-Ts. This is the type of E-T on which my research project concentrates.

Transitional Environmental Technologies

Contemporary E-Ts must be global and permanent applications. Regional or temporary E-Ts such as pollution treatment of production, fossil-fuel-powered electric cars, cleaner fossil fuels, air or water filtration systems, end-of-pipe technologies are regarded as only transitional measures for E-Ts. The concept of E-Ts is therefore a broad and long vision; otherwise the situation will fall into a paradox that is fraught with the problems of overconsumption and environmental pollution.

Mature and Infant Environmental Technologies

Mature E-Ts range from the technique of highly efficient cogeneration systems of production, renewable energy in a mechanical form, forest management, cultivation and land management, materials recycling, and innovative waste disposal. Although most of the functions and features in these technologies are environmentally conservative, the movable parts, consumables, and the processes involved in materials transformation create inefficiency and this is largely inevitable. Nevertheless, most of the mature E-Ts nowadays are the most practical solution for sustainable development. How to further transform them into purely pollution-free technologies will be a matter for further development of the form, size, and dimension of existence and application of E-Ts (e.g., the online visual convention and remote control of production and other activities may reduce much of the polluting consumption). Nuclear energy, genetic engineering, petrochemical products, and so forth arouse argument because they do not pollute environment but are nonrestorable or disturb the ecosystem in the long run. The environmental application of biotechnology and new material technology such as liquefied nitrogen superconductors, superdurable compound materials, super magnetic nanomaterials are still in the R&D phase and for this reason I exclude these infant or future E-Ts from the focus of my study and choose three of the most popular and mature E-Ts at the moment, that is, wind energy, packaging waste recycling, and environmentally friendly auto technology.

Resources Management of Environmental Technologies

In addition, since the concept of resources management has been incorporated, the mode of technology development will now tend to shift from centralized technologies to distributed ones (e.g., central power supply to distributed power supplies) and from the real world to the virtual world (e.g., real meetings to Internet meetings). Centralized technologies such as large hydro or nuclear power plants are subject to breakup and debacle. They also tend to be inefficient, wasteful, and harmful to the environment, whereas distributed technologies are flexible, efficient, and, most importantly, benign in their environmental effects.

APPENDIX II   TYPES OF ENVIRONMENTAL TECHNOLOGY POLICIES IN MARKET ECONOMIES

Evolution of Environmental-Technology Policies

Tisdell (1979: 25) states that "technology policy is concerned with the adoption and use of techniques—the innovation, diffusion of techniques and their replacement." Accordingly, E-T policy is also a plan of action in response to the development or economic problems of E-Ts. It is designed to achieve goals and is a formally legislated expression of intent to achieve goals through a course of programs within some specified time. Over the past few decades, many governments have begun to recognize the importance of ecological changes and resource scarcities. Ozone depletion, rising sea-levels, and global warming caused by CO_2 emission are surely harmful to life on earth. The damage to the global ecosystems of the air, land, water, and living species is also increasingly becoming a factor of tensions and conflicts among people and countries. Governments have to react to these issues. In addition, environmental problems begin to adversely influence the economic growth ranging from tourism, fishing, the petrochemical and automobile industries, and food production. Thus, while economic and security concerns are traditionally given priority among policymakers, making E-T policies is now also an important agenda of public policy discussion, wherein some of the E-T policies are designed to promote new technologies to resolve the environmental problems and the E-T development issues, and to promote especially those that may economically replace the polluting technologies or activities.

Range of E-T Policies and Policy Instruments

Although E-T policies are only a part of a wider environmental policy, they include a wide range of measures and actions to advocate new technologies usually within some specified period. Policymakers may employ programs to create public concern for E-Ts through education, promotion, exhibition and propaganda services, consumer protection and incentive, and mandatory consumption. To the E-T industry and other E-T policy actors, policymakers may provide them with preferential conditions such as easy access to grants, capital or low-interest financial assistance, special amortization benefits, R&D subsidies, tax exemption or privilege, preferential duties on imported materials and equipment, licensing of imported foreign components, less import restrictions, providing infrastructure (such as assistance with research, development, and demonstration), sales and production facilities, expertise and information archives and transportation means. To encourage corporations to adopt E-Ts, environmental standards, regulatory regimes, innovation waivers, eco-taxes, tradable permits, covenants, investment subsidies, eco-labels, environmental management and auditing systems, game management, network management, societal debates, technology foresight, public/private partnerships, voluntary agreements, goal setting and indicative planning, innovative clusters and strategic niche management can be applied either to leverage private funds for market learning investment or to stimulate dynamic efficiency and flexibility for better market forces (see table A.1). Still, some E-T policies are based on fair rules and regulations, or on effective and fair competition, to impose taxes or charges on pollutants and polluting activities.

Objectives and Stages of Environmental-Technology Policies

The specific strategies and programs of E-T policies are to promote new technologies in two categories: first, to reduce environmental damage and to prevent new pollution. New technologies such as those that amend the depletion of the ozone layer or reduce the greenhouse, effect E-Ts that have no commercial value, have to be subsidized by governments; second, to enhance demand and to increase competitiveness, technologies such as solar cells or recycling appliances that have a high market potential have to be incubated until the industry manufacturing such products is able to design, produce, and distribute products and services at costs that are market-competitive.

The objectives of E-T policy have different stages. Some of them are designed to help research and development (R&D) and to stimulate innovation. Others are targeted at public deployment or demonstration (D) to let E-T actors absorb experience in the market in order to reach commercial maturity. Many governments' early E-T programs are therefore called RD&D. Others are focused on how to attract private investment in order to minimize government expenditures. The ultimate target is to terminate government support and to fully privatize and diffuse the E-Ts.

Table A.1 Policy Instruments to Promote the Development and Use of Environmentally Beneficial Technologies in Different Contexts

Policy instrument	General inherent characteristics	Purpose for which they may be used	Context in which they may be applied
Technology-based environmental standards	* Effective in most cases (when adequately enforced) * Uniform standards give rise to inefficiencies in case of heterogeneous polluters	Technological diffusion and incremental innovation	When differences in the marginal costs of pollution abatement are small and economically feasible solutions to environmental problems are available
Technology-forcing standards	* Effective (in focusing the attention of industry on environmental problems) * Danger of forcing industry to invest in overly expensive and suboptimal technologies * Problem of credibility	Technological innovation	* When technological opportunities are available that can be developed at sufficiently low cost * When there is a consensus about the appropriate compliance technology
Innovation waivers	Same as technology-forcing standards	Technological Innovation	When technological opportunities are available and when there is uncertainty about the best solution
Eco-taxes	* Efficient * Uncertainty about industry response * Danger that they provide a stimulus which is too weak and indirect * Total environmental costs for industry are likely to be high * Limited political attractiveness	* For recycling and material and energy saving * Technological diffusion and incremental innovation	* In case of heterogeneous polluters that respond to price signals * When there are many different technologies for achieving environmental benefits

Tradable permits	* Effective * Cost effective (which means that environmental benefits are achieved at lowest cost)	Technological innovation and diffusion	* Same as for taxes * Costs of monitoring and transaction should not be prohibitively high
Covenants and technology compacts	* Uncertainty about whether industry will meet agreements; should be supplemented with penalty for noncompliance * Low administrative costs	Technological diffusion	* In case of many polluters and many technological solutions * When monitoring environmental performance is expensive
R&D subsidies	* Danger of funding second-rate projects * Danger of providing windfall gains to recipients	Technological innovation	* When markets for environmental technology do not yet exist and when there is uncertainty about future policies * When there are problems of appropriating the benefits from innovation * When there are important knowledge spillovers * In case of large social benefit and insufficient private benefits
Investment subsidies	* In conflict with polluter pays principle * Danger of windfall gains * Politically expedient	Technological diffusion	When industry suffers a competitive disadvantage due to less strict regulations in other countries

Continued

Table A.1 Continued

Policy instrument	General inherent characteristics	Purpose for which they may be used	Context in which they may be applied
Communication (e.g., eco-labels)	* Helps to focus the attention of firms and consumers on environmental problems and available solutions to these problems * Little coercive power	Technological diffusion	* When there is a lack of environmental consciousness * When there are information failures
Environmental management and auditing systems (EMAS)	* Enhance environmental knowledge and competence * Little coercive power	Technological diffusion, product improvement and good housekeeping	In case of lack of environmental knowledge and competence
Network Management	* Create a platform for learning and interaction, to stimulate alignment and coordinate interdependent activities. Solutions may be tailored to specific needs * Requires technological understanding of processes and products	Technological diffusion and innovation	When there are information failures
Societal debates on environmental issues		* For stimulating mutual understanding and learning about values and belief systems * For improving processes of anticipation	In case of controversies over problems and solutions

Sustainability foresight studies	* Broadens processes of assessment * Enhances strategic orientation	* For learning about sustainability options (beyond eco-efficiency) * For altering fixed ideas and mind sets	
Setting of goals and use of indicative planning	Provides clarity and (strategic) orientation	For shaping business expectations and guiding strategic decisions	
Game management		Radical innovations with significant sustainability benefits that do not offer a win-win solution	In case of oligopolies engaged in strategic behavior over environmental issues
Strategic niche management		For learning about radical innovations and for stimulating processes of coevolution	* For pathway technologies to a more sustainable system * In case of attractive domains of application

Source: Innovation and the Environment, OECD, 2000, pp. 52–54.

The Making of Environmental-Technology Policies

The formation of policies is subject to how and where to investigate the problems, in what areas the technologies can be commercialized (or for environmental protection only), how and who designs the policies, how and who pays the necessary expenses of policy implementation, how and who implements them, how and when to control and measure the process as well as the outcomes of stages. Governments in different countries and levels have different methods of policymaking. For example, American federal government policymaking (except those involving national security) has to take place through constitutional processes that is publicized for study and debate, while Japanese policymaking does not rely on lucidly self-explained laws but resorts to the expertise and intelligence of the bureaucracy interpreting laws at their discretion and delivering administrative guidance to the industry (Nester 1990: 138). These have formed the respective basic principles and procedures for the formulation of E-T policies in these countries, on which the role of governments has exerted the greatest impact.

Notes

1   Traditonal Institutions are Ineffective in an Exceptional Policy Sector

1. According to OECD definitions, market failures "may arise when firms lack information about new or established technologies or are disadvantaged because of size or scale production requirements" and systemic failures "arise from weakness in linkages and interactions among different actors in national innovation systems" (1997: 11).
2. See Appendix 1.1, policy instruments to promote the development and use of environmentally beneficial technologies in different contexts.
3. See more from O'Neill, Kate, *Waste Trading Among Rich Nations*, The MIT Press, Massachusetts, 2000, pp. 162–163.
4. The data were obtained from the American Aluminum Association.
5. The data are obtained from the American Plastics Council.
6. The data are obtained from the U.S. Bureau of the Census.
7. The comment is sourced from CEI News Center at <http://www.cei.org/gencon/003,02613.cfm> searched on July 28, 2004.
8. More details can be obtained from the basic charts that document the past few years of Vestas (and NEG Micron) at <http://uk.finance.yahoo.com/q/bc?s=>VWS.CO&t=5y> searched on February 6, 2006. The former acquired the latter in 2004.
9. Sourced from the Institute for Sustainable Energy Policies at <http://www.isep.or.jp/e/Eng_project.html> searched on October 6, 2003.
10. Sourced from the EV World at <http://evworld.com/archives/conferences/futurecar2000/jcooper.html> searched on June 24, 2004.

2   Tweaking Institutions in the Shadow of the State

1. The market forces versus state powers debate started when Adam Smith in the mid 1770s published his seminal work on capitalism and liberalism: *The Wealth of Nations*. Reference can be made to Adam Smith's *An*

Enquiry into the Nature and Causes of the Wealth of Nation, edited by Edwin Cannan (London: Methuen, 1961).

2. Sourced from the Prepared Witness Testimony by Mr. Jeff Welton, senior vice president, Wintec Energy to the Subcommittee on Energy and Air Quality on January 12, 2004 at 10:00 AM, Civic Center Council Chamber, available at <http://energycommerce.house.gov/108/Hearings/01122004hearing1162/Welton1829.htm> searched on February 7, 2006.

3. There are also heated debates over the relative value of the "calculus approach," the "cultural approach," and their amalgamation (Hall and Taylor 1996; Hay and Wincott 1998). This book is based on an assumption that both of them have an impact on the reaction of actors to E-T policies.

3   "DIFFERENT TRAITS DESIGN" COMPARISONS

1. According to the report of American Wind Energy Association, during 1999, new wind power units with a capacity of over 3,600 MW was installed globally, cumulating a total capacity of 13,400 MW. These figures imply an over 36% increase since 1998, during which installation totaled 9,751 MW.

2. In the 1980s, Japan was the world's largest auto producer (e.g., 12.7 million units or 26% of market share in 1988), whereas the United States was the second largest (e.g., 11.2 million or 23%). The third and fourth largest auto producers were Germany (e.g., 3.9 million in 1988) and France (e.g., 3.4 million in 1988) respectively, but they had far less production units than Japan and the United States. However, the downturn of Japan's domestic sales and the surge of the U.S. auto market since the 1990s have made the latter became the world's largest auto producer. The latest shrinkage in Japan auto production can be found on the Web site of Japan Automobile Manufacturers Association <http://www.jama.org/statistics/motorvehicle/production/mv_prod_month.htm>.

4   VARIETIES OF REGULATORY STATE

1. A modified version of this chapter titled "Obliging Institutions and Industry Evolution: A Comparative Study of the German and U.K. Wind Energy Industries" was published in *Industry and Innovation*, Vol. 12, No. 1 (March 2005); and some ideas of this chapter are sourced from "The Telecommunications Regulatory Regimes in Hong Kong and Singapore: When Direct State Intervention Meets Indirect Policy Instruments," coauthored by Painter, Martin, and Shiu-Fai Wong, paper delivered to the Fourth International Convention of Asian Scholars, Shanghai Academy of Social Sciences, August 20–24, 2005.

2. Wind power technology was very primitive in the early 1990s but the following decade saw major developmental progress, wind power is now

cheaper than coal or nuclear energy and is starting to challenge the competitiveness of gas energy.

3. Data has been obtained from the European Wind Energy Association at <http://www.ewea.org/doc/03-03-03%20World%20figures.pdf>.

4. While quite a few developed countries have declared their opposition to the 1997 Kyoto Protocol, the seventh Conference of the Parties to the UN Framework Convention on Climate Change adopted the Marrakesh Accords set out the way of implementing the protocol in detail in November 2001 and allowed domestic actions to be taken in 2005 for a target of reduction of CO_2 emissions by 10% by 2010.

5. The European Economic Community (EEC) was formed in 1957 to remove trade barriers between its member states and to form a common market. In 1967 the EEC and the other two European communities were merged leading to the establishment of a single commission, a council of ministers, and a European Parliament.

6. Out of the 15 member states of the EU, only Germany, Finland, Sweden, and the United Kingdom were fully deregulated as soon as the European directive took effect in 1998.

7. More detailed requirements of latest European Electricity market liberalization can be found at the EU Web site <http://europa.eu.int/comm/ energy/electricity/index_en.htm> and <http://europa.eu.int/ comm/ energy/electricity/ legislation/ index_en.htm>.

8. According to the report of the American Wind Energy Association, during 1999, new wind power units with a capacity of over 3,600 MW was installed globally, cumulating a total capacity of 13,400 MW. These figures imply an over 36% increase over the 1998 total instalation figure of 9,751 MW.

9. The United Kingdom was always a coal exporter, particularly during its peak production period in the early twentieth century. However, since the early 1980s, Britain has imported much more coal than it has exported.

10. More information about the "Task Force" and "Pilot" can be found at <http://www.red-star-research.org.uk/subframe2.html>.

11. The Department of the Environment, Transport and the Regions (DETR) was later split into the Department for Transport, Local Government and the Regions (DTLR), and the Department for Environment Food and Rural Affairs (DEFRA) in June 2001. The DTLR has been further replaced by the Department for Transport (DfT) by the decision of the British prime minister in May 2002, while the DEFRA has recently taken over the functions of the Ministry of Agriculture, Fisheries and Food (MAFF), following the disastrous handling of the foot and mouth disease pandemic and its widespread condemnation.

12. Wind power developers produced and sold electricity to the "pool" administrated by the national grid through the NFFO/SRO tendering system.

13. The Office of Electricity Regulation (OFFER) was merged later in January 1999 with the gas market regulator Office of Gas Supply (OFGAS) to become a new common regulator Office of Gas and Electricity Markets (OFGEM) empowered by the Utilities Act 2000.

14. More information about NFPA can be found at <http:// www. nfpa.co. uk/ id17n.htm>.

15. The selling price of wind power during the first round of the Non-Fossil Fuel Obligation (NFFO-1) was over nine pence per kWh, but was 2.8 pence per kWh by 1998, and two pence/kWh in Scotland during the third round under the Scottish Renewable Obligation (SRO).

16. Conventional electricity was sold to the "pool" through a daily bidding system. The pool price varied every half an hour according to actual supply and demand. Consumers entered contracts with their suppliers for a stable buying price, and consumption was charged by remote measurement via telephone lines.

17. More details about the stories given here as examples can be found from the Web site of Country Guardian available at <www.countryguardian. net/cefncroes.htm>.

18. Sainsbury's Distribution Depot in Scotland was the first wind farm built and run under the Merchant Wind Power Initiative 2001 by a British energy company on March 19, 2001, taking three days to complete the erection and commissioning and to supply wind energy with subsidy and premium. The Ecotricity's second Merchant Wind Power initiative was built at Ford's Dagenham plant in London in August 2003.

19. Renewable Energy Systems group (RES) has been one of the world's leading IPPs for twenty years and has energy projects with over 700 MW of wind energy capacity across Europe and North America.

20. More information about BMBF can be found at <http:// www.BMBF. de>.

21. Third Reich refers to the Nazi regime in Germany between 1933 and 1945, during which the Nazi government had guaranteed the utilities a monopolized market to assure the supply of electricity.

22. More information about the International Energy Agency (IEA) can be found at <http://www.iea.org>.

23. English language charts and data from the WMEP program can be found at <http://reisi.iset.uni-kassel.de/wind/reisi_dw.html>.

24. More information about BWE can be found at <www.wind-energie. de>.

25. More information about BMWA can be found at <http://www. bmwi.de> (the name of bmwi remains unchanged on the Web address).

5   Varieties of State-Societal Structure

1. Packaging waste recycling is usually coupled with waste recovery in governmental waste management policies. Yet recycling treatment refers

more to the new packaging that is recyclable (e.g., melting and refabricating plastic bottles), rather than to recoverable packaging (e.g., cleaning and pasteurizing glass bottles).

2. Container Deposit Legislation refers to the regulation of a *mandatory* deposit on containers to encourage their return by consumers, while Curbside Recycling System refers to separate bins for recyclable materials from household consumers.

3. Coalition government is a cabinet of government in which several political parties cooperate.

4. The Federal Trade Commission enforces federal consumer protection laws and federal antitrust laws that prevent fraud, deception, and unfair business practices and prohibit anticompetitive mergers and other business activities that constrain competition and put consumers at a disadvantage.

5. The New Federalism in the United States is a policy theme of the 1980s and 1990s. In addition, a block grant is a large sum of revenue returned by the federal government to a state government with only general conditions of spending.

6. The information is sourced from Container Recycling Institute at Web sites <http://www.Container-Recycling.org> and <http://www.Bottlebill.info> (November 22, 2002).

7. Sourced from the Container and Packaging Recycling Updates and the short article "Missouri: New Attack on Local Deposit Ordinance" available on the Web site of the Container Recycling Institute, <http://72.14.207.104/search?q=cache:WDVGY8TKpyIJ:www.container-recycling.org/assets/pdfs/newsletters/CRI-NL-2001SummerFall.pdf'Columbians'Against'Throwaways'In'1976,'&hl=en&gl=hk&ct=clnk&cd=6> (December 6, 2003).

8. According to the Statutory Instrument 1998 No. 1762 (N.I. 16), the "producer responsibility obligation" refers to "the steps which are required to be taken by relevant persons of the classes or descriptions to which the regulations in question apply in order to secure attainment of the targets specified or described in the regulations." More details can be found from the Web site of the Office of the Public Sector Information, available at <http:// www.opsi.gov.uk/si/si1998/19981762.htm>. Meanwhile the purchase of "packaging waste recovery notes" is the act of acquiring documentary evidence to prove the fulfillment of the packaging waste producer responsibility.

9. The United Kingdom has a devolved system of government in which subgovernment bodies or regional parliaments are established but exist only by virtue of statute or ordinary law rather than constitutional law, and they can be abolished or have their functions altered within the discretion of the central parliament. Usually the devolution is confined to financial or political issues.

10. The information was obtained from the essay "The Household Waste Recycling Bill has passed its next test on the way to becoming UK law," by James Cartledge, available at http://www.letsrecycle.com/materials/packaging/news.jsp?story=2418 (July 11, 2003).

11. The data are obtained according to Australia's House of Representatives Fact Sheet No. 6. In addition, a private member is any parliamentary member other than the executive members such as prime minister, the speaker, a minister or parliamentary secretary.

12. The Beverage Industry Environment Council is Australia's leading industry association representing the environmental interests of Australia's drink manufacturers and their aluminum, glass, and PET packaging suppliers for more than two decades. More information can be found at <http://www.biec.com.au/index.html>. The Australian Council of Recyclers is Australia's leading industry association, representing solid waste recyclers since 1983. More information can be found at <http://www.acor.org.au/whoisacor.html>.

13. More information about DSD can be found at < http://www.gruener-punkt.de>.

14. More information can be found from my source *Waste Trading Among Rich Nations*, by Kate O'Neill (Cambridge, MA: The MIT Press, 2000), pp. 162–163.

15. More information on Article 20a of the Basic Law can be found at <http://www.iuscomp.org/gla/statutes/GG.htm>.

16. More information of *AB Svenska Returpack* can be found at <http://www.returpack.se/>.

17. More information of *Svensk GlasÅtervinning AB* can be found at <http://www.glasbanken.com>.

18. One Swedish öre is equal to a one-hundredth subdivision of the Swedish krona.

19. More information of Swedish Environmental Protection Agency can be found at <http://www.internat.environ.se>.

20. The official Web sites of the IGORA and the PRS are respectively <http://www.igora.ch/> and http://www.petrecycling.ch/all/frameset.cfm?vDom=1 (July 18, 2003).

21. The board members of the PRS in 2003 included Henniez SA, Coca-Cola Beverages AG, Evian-Volvic Suisse SA, *Feldschlösschen Getränke AG*, and Rivella AG.

6 Varieties of State-Societal Relationship

1. More information about oil consumption by the transport sector can be found in *The IEA and Transport* at <http://www. worldenergyoutlook.org/> (December 2, 2003).

2. Environmentally friendly auto technology is a general term referring to high efficiency, low-polluting vehicle technologies and fuels. They may include lean-burn combustion, liquefied petroleum gas, compressed natural gas, electric, hybrid, fuel-cell cars, and the like.

3. Overcapacity of car production has also been a side problem of environmentally friendly car development. Automakers around the world in 2003 had a production capacity of about 77 million cars and light trucks,

registering an increase of 19 million units since 1990, according to Autofacts, a PricewaterhouseCoopers subsidiary in Bloomfield Hills, Michigan. But consumption is estimated to total only about 56 million units.

4. Auto exhausts are one of the most serious sources of CO_2 emission levels. For example, in 1996, it was estimated that 6 billion tons of carbon was discharged into atmosphere annually and that the United States has been the largest polluter in this sector, followed by Japan.

5. More information about USCAR can be found at its Web site <http://www.uscar.org>.

6. On November 19, 2002, Toyota announced that a new market-ready fuel cell hybrid vehicle developed from the Prius gas-electric hybrid technology would be available for leasing from December 2, 2002 onward. Success in this area was enjoyed not only by Toyota, other Japanese automakers soon followed suit to offer their clean-fuel vehicles. Honda launched the Insight hybrid in January 1999 and Nissan came next with the Tino hybrid in April 2000 (though this hybrid model was soon discontinued). In addition to the specification that gas-electric hybrid cars reduce fuel consumption by around 50% and emit emissions 90% below the current standard, Prius and Insights/Civic hybrids are totally transformable into zero emission fuel such as fuel cells. General Motors was the first American automaker to announce commercialization of its hybrid design but the dream was promised to come true only by 2007. In the early 2000s, the four-seat Toyota Prius and the two-seat Honda Insight were the only gas-electric hybrids available on the market, while eight new models were launched around 2004. Toyota sold some 20,000 Prius hybrids in United States in 2002, an increase of 29% from 2001, and Honda also sold a similar number of its Insight/Civic hybrids during the same period. Up to the North American International Auto Show at Detroit, Michigan, in January 2004, virtually all the top EFA models available in showrooms for sale were Japanese.

7. California was the first American state to launch tight emission standards. As part of its 1990 program, 10% of newly registered vehicles were to have zero emissions by 2003. However, following a compromise made by the government, the automakers were allowed to meet a lower standard, i.e., Partial Zero Emission Vehicle (PZEV).

8. Some more of the stories and other information can be found in the papers written by Kemp R. (2002).

9. While the U.S. retail gasoline prices were at a low level (rose from 122.2 cents/gallon on December 26, 1994 to 142.3 on November 25, 1996, and then went down to 107.4 on February 22, 1999), the Japanese prices were around two times the U.S. prices because of its high domestic tax. In January 2003 the American price was 145.3 cents/gallon but the Japanese one was 334.8 cents/gallon.

10. The National Research Council (NRC) is a study affiliate of the National Academy of Sciences. Since 1994, NRC has set up a regular committee

(funded by government sponsors of PNGV) for evaluating the progress of PNGV.

11. In the 1980s, Japan was the world's largest auto producer (e.g., 12.7 million units or 26% of market share in 1988), whereas the United States was the second largest (e.g., 11.2 million or 23%). The third and fourth largest auto producers were respectively Germany (e.g., 3.9 million in 1988) and France (e.g., 3.4 million in 1988) but they had far less production units than Japan and the United States. However, the downturn of Japan's domestic sales and the surge of the U.S. auto market since the 1990s have made the latter became the world's largest auto producer. Information on the latest shrinkage in Japan auto production can be found at the Web site of Japan Automobile Manufacturers Association at <http://www.jama.org/statistics/motorvehicle/production/mv_prod_month.htm> (December 9, 2003).

12. More information about METI can be found at <http://www.meti.go.jp/> (December 9, 2003).

13. Information obtained from the article, "Smelly Business: Car Makers and Air Pollution," by Ralph Nader, June 9, 1999 at <http://www.nader.org/interest/060999.html> (December 9, 2003).

14. Part of the material of this section is drawn primarily from the article, "Wheels of Change: Tax Policy Effects on U.S.-Japan Auto Trade," by Akinori Tomohara (2002), and a draft paper presented at an East Asian Institute seminar (which is cosponsored by the Center on Japanese Economy and Business) at Columbia University, held on February 12, 2002.

15. U.S.-based Japanese automakers include Honda in Ohio, Toyota in Kentucky, and Nissan in Tennessee.

16. The Corporate Average Fuel Economy (CAFE) program was firstly devised in 1975 as soon as the U.S. Congress passed the Energy Policy and Conservation Act, aiming at reducing the reliance on foreign oil after the 1973 Oil Crisis. The CAFE required the automakers to increase the average fuel economy standard of passenger cars from 18 miles per gallon (mpg) in model year 1978 to 27.5 mpg in model year 1985 and the standard of light trucks including minivans, pick-ups, and sport utility vehicles to 20.7 mpg in model year 2002. The California State in 1990 mandated that 2% of vehicles sold in 1998 (5% in 2001 and 10% in 2003) by any automaker must fulfilll the ZEV mandate, i.e., zero emitting.

17. Quotation from Mathew Wald, "Government Dream Car," *The New York Times*, September 30, 1993, p. 1.

18. The PNGV's goals include (1) significant improvement of national competitiveness in auto manufacturing; (2) commercialization of innovations on conventional vehicles; (3) development of efficient, environmentally friendly autos, also fulfilling market's requirements for safety, quality, performance, utility, and affordability.

19. The PNGV participants from the government side were DOC, DOE, the Department of Defense (DOD), the Department of Transport (DOT),

the Environmental Protection Agency (EPA), the National Aeronautics and Space Administration (NASA), the National Science Foundation (NSF), and executive offices of the president as well as other designated universities, laboratories, and even suppliers.

20. More information about PNGV reviews from Technology Administration, Department of Commerce can be found at <http://www.ta.doc.gov/pngv/news/PeerReviews.htm> (December 20, 2003).

21. Quotation found in the article, "The Partnership for a New Generation of Vehicles Is Corporate Welfare At It's Worst," by Ralph Nader in *San Francisco Bay Guardian*, June 19, 2000.

22. Sourced from the online news article "Powerful, Super Clean Focus Helps Ford Meet Its Environmental Commitments" on September 10, 2003 in the AutoFan Magazine, available at http://www.autofan. com/newsdetail.asp?id=664&mn=10&yr=2003> (March 24, 2004).

23. CFO magazine is a specialist finance magazine edited for senior financial executives.

24. The annual NRC reports can be found on the General Accounting Office at <http.//www.gao.gov> (March 24, 2004).

25. More information about the DOE's FreedomCAR Program can be found at <http://www.cartech.doe.gov/freedomcar> (March 24, 2004).

26. More information about OTT can be found at its Web site <http://www.ott.doe.gov> (March 24, 2004).

27. The United States had also provided the car industry (except for the PNGV) with a 20% tax credit for the R&D portion since 1980 and renewed it ten times until its expiration in June 1998.

28. MTBE stands for methyl tertiary-butyl ether, it was reported to be a gasoline additive that was contaminating groundwater, causing many citizens from various cities and states in the United States to file lawsuits. Some more information can be found from the Statement of Scott H. Segal Bracewell & Patterson, L.L.P. Counsel, Oxygenated Fuels Association before the Subcommittee on Clean Air, Climate Change and Nuclear Safety, Committee on Environment and Public Works, United States Senate, the Hearing on Proposals Regarding Fuel Additives and Renewable Fuels on March 20, 2003.

29. MITI (Ministry of International Trade and Industry) is the former name of METI (Ministry of Economy, Trade and Industry).

30. More information about NEDO can be found at its Web site <http://www.nedo.go.jp/english/introducing/activities/ newenergy/ conservation.pdf#d> (March 24, 2004).

31. There are stringent EU requirements, e.g., Euro Three and Euro Four for lower-emission engines introduced in 2000 and 2005 respectively. They specify so high requirements on fuel quality that the auto and relevant industries have to improve their engine designs, the process of auto production (in factories), and distribution of fuel gas (through filling stations).

32. Part of the information was obtained from an interview with Mark Amstock, the Toyota Advanced Technology Product Launch Manager, conducted by Bill Morre on September 29, 1999 for his essay "Behind the Scenes of Prius Launch," at <http://www.evworld.com/archives/interviews2/amstock.html> (March 31, 2004).

33. Sourced from the Toyota's news release on November 25, 2005, available at <http://www.toyota.co.jp/en/news/05/1125_2.html> (February 15, 2006).

34. Part of the information was obtained from an interview with Honda's CEO, Hiroyuki Yoshino, conducted on December 23, 2002 for the essay "Honda's CEO on Fuel Cells' Future," in *Business Week* available at <http://www.businessweek.com/magazine/content/02_51/b3813085.htm> and U.S. Department of Energy at <http://www.eere.energy.gov/vehiclesandfuels/facts/2003/fcvt_fotw251.shtml>.

35. Sourced from Honda's Annual Report (year ended March 31, 2005), available at <http://world.honda.com/investors/annualreport/2005/pdf/ar2005.pdf> (February 15, 2006).

36. Details about the reasons for the popularity of SUVs can be found from the paper, "Analysis of the Impact of Sport Utility Vehicles in the United States," by Stacy C. Davis and Lorena F. Truett in August 2000, prepared for the Office of Transportation Technologies at the U.S. Department of Energy.

37. The data can be found from the annual reports of Toyota.

38. The data can be found from the annual report of GM.

39. Sourced from the cover story of the February 1996 edition of *Newsweek* featuring an essay called "Corporate Killers" via Web site <http://www.newwork.com/Pages/News/NewsArchives/1996/Feb96_news.html> (June 12, 2003).

40. Sourced from "CEO Bill Ford Jr. Takes on his Critics: He Dismisses Reluctant Leader Label," by Mark Truby for *The Detroit News* Website <http://www.detnews.com/2003/autosinsider/0306/13/a01-192223. htm> (June 13, 2003).

41. Sourced from "Smelly Business: Car Makers and Air Pollution," by Ralph Nader on June 9, 1999 at <http://www.nader.org/interest/060999.html> (June 13, 2003).

42. Nader, loc.cit.

43. Since the Energy Policy and Conservation Act of 1975 established corporate average fuel economy (CAFE) standards for new passenger cars, the American auto industry has been struggling to meet the increasing requirements of CAFE and petitioning to stay from being penalized for noncompliance.

44. "Cynics think that the PNGV was simply a politically astute 10-year reprieve for the domestic auto industry from threats of higher Corporate Average Fuel Efficiency standards," Earth Day founder Denis Hayes in his new book, *The Official Earth Day Guide to Planet Repair*. (Washington, D.C.: Island Press, 2000).

45. The LDP has continuously been in power since 1955 (except for 1993), the present prime minster, Junichiro Koizumi, is a member of the LDP. Thus, a LDP-dominated alliance in the Diet has a long history.

46. The statement is quoted according to the explanation given in an interview with Hiroyuki Yoshino, the CEO of Honda, in *Business Week* in December 2002

47. More information about the IEA Hybrid & Electric Vehicle (IEAHEV) Implementing Agreement can be found at <http://www.ieahev.org> and about the Hybrid Vehicles Overview Report 2000, IEA, at <http://www.ieahev.org/pdf/05-Annex-VII00.pdf> (June 13, 2003).

48. The National Academy of Sciences is a private, nonprofit society of distinguished scientific and engineering scholars, having a mandate since 1863 to advise the federal government on scientific and technical issues. The National Academy of Engineering was set up in 1964, as a branch of the National Academy of Sciences

49. The National Research Council, *Effectiveness of the United States Advanced Battery Consortium as a Government-Industry Partnership* (Washington, DC: National Academy Press, 1998), p. 60.

50. The Japanese government helps standardize design and production of parts, especially when the competitiveness of smaller automakers becomes low. For example, in 1994, the Ministry of International Trade and Industry standardized 84 parts in cooperation with six manufacturers of small autos (Nikkei Sangyo, January 6, 1994).

51. As compared to the nationwide tier I emissions standard, the SULEV standard requires 97% less hydrocarbon emissions, 76% fewer carbon monoxide, and 97% less nitrogen oxide; and therefore diesel engine normally does not fulfill the low carbon monoxide requirement.

52. The comment was made by William Safire who has for the past quarter century been the editor for *The New York Times*. He is the winner of 1978 Pulitzer Prize for distinguished commentary and was also a White House advisor and speechwriter in the Nixon administration.

7   CONCLUSIONS

1. The revisionist view was an opinion in the anti-Marxist socialist movement arguing against revolutionary Marxist theory and promoting peaceful reforms for social progress.

2. More information can be found in the article "High-Speed Chase," by Assif Shameen and Mutsuko Murakami in *Asia Week* on July 6, 2001, available at <http://www.pathfinder.com/asiaweek/magazine/nations/0,8782,165855,00.html> (June 20, 2003).

3. Arguments about neoutilitarianism that came in later can be found in the comparison that Yan (1998) makes between western and Chinese utilitarianisms, leading to a new thought of neoutilitarianism and the Ruggie's (1999) description of neoutilitarianism, containing realism and neoliberalism.

BIBLIOGRAPHY

AB *Svenska Returpack*, aluminum can and PET bottle recycling rates in various years, <http://www.returpack.se/> (September 30, 2003).

ACI Glass Packaging, recycling glass packaging rates 2000, <http://www.acipackaging.com/aciwww.nsf/FrameSet?OpenFrameset> (September 30, 2003).

Aghion, Philippe, and Peter Howitt. 1998. *Endogenous Growth Theory*. Cambridge, MA: MIT Press.

Aluminium Can Group, Australia, aluminum can recycling rates 2000, <http://www.aluminium-cans.com.au/> (October 2, 2003).

Aluminum Association Inc., aluminum can recycling rates in various years, <http://www.aluminum.org/> (October 20, 2003).

American Plastic Council, PET bottle recycling rates in various years, <http://www.americanplasticscouncil.org/s_apc/index.asp> (October 2, 2003).

American Wind Energy Association, *Global Wind Energy Market Report 2003*, <http://www.awea.org/> (February 2, 2004).

———, "How Much Funding is Provided for Federal Wind Energy Research and Development?" <http://www.awea.org/faq/fedrdfnd.html> (April 15, 2004).

———, "How Much Funding is provided for Federal Wind Energy Research and Development?" <http://www.awea.org/faq/fedrdfnd.html> (September 25, 2003).

Ancona, D.F., P.R. Goldman, and R.W. Thresher. 1996. "Wind Program Technological Developments in the United States." Paper presented at the World Renewable Energy Congress on June 18, Denver, CO.

Anderson, Evelyn. 2003. "Toyota and Denso—A Rarely Researched Aspect of Japan's Competitiveness." Paper presented at the 13th Biennial Conference of the Japanese Studies Association of Australia, July 2–4, held at the Queensland University of Technology, Brisbane, Australia.

Arter, David. 1990. The Swedish Risksdag: The Case of a Strong Assembly. In *Parliaments in Western Europe*. Edited by Norton Philip. London and Oregon: Frank Cass.

Aumônier, S., and J. Troni. 2000. *Research Study on International Recycling Experience*. Report prepared by Environmental Resources Management for the Department for Environment, Food & Rural Affairs of the U.K.

Barberis, Peter, and May Timothy. 1993. *Government, Industry and Political Economy*, Buckingham: Open University Press.

Barrett, Brendan, ed. 2005. *Ecological Modernisation and Japan*. London: Routledge.

Barro, Robert J., and Xavier Sala-i-Martin. 1995. *Economic Growth*. New York: McGraw-Hill.

Bell, Stephen, and John Wanna, eds. 1992. *Business-Government Relations in Australia*. Sydney, Fort Worth: Harcourt Brace Jovanovich Publishers.

Bell, Stephen, ed. 2002. *Economic Governance and Institutional Dynamics*. South Melbourne: Oxford University Press.

Bergek, Anna. 2002. "Shaping and Exploiting Technological Opportunities: The Case of Renewable Energy Technology in Sweden," Ph.D. Thesis, Department of Industrial Dynamics, Chalmers University of Technology Göteborg, Sweden.

Beverage Industry Environment Council, "Scoping Paper for Beer and Carbonated Soft Drink IWRP," <http://www.biec.com.au/> (October 4, 2003).

Beynon, R., ed. 1999. *The Icon Critical Dictionary of Global Economics*. Cambridge, U.K.: Icon Books Ltd.

BioCycle, bottle recycling rates in various years, <http://www.jgpress.com/biocycle.htm> (October 4, 2003).

Bobrow, Davis B., and John S. Dryzek. 1987. *Policy Analysis by Design*. Pittsburgh: University of Pittsburgh Press.

Bonoli, G. 1999. *State Structures and the Process of Welfare State Adaptation*, Department of Social and Policy Sciences, University of Bath, Mimeograph.

Borrus, Michael, François Bar, Patrick Cogez, Anne Brit Thoresen, Ibrahim Warde, and Aki Yoshikawa. 1985. Telecommunications Development in Comparative Perspective: The New Telecommunications in Europe, Japan, and the U.S. Working Paper No. 14 at the Berkeley Roundtable on the International Economy (BRIE), May.

Branscomb, Lewis, Florida, Richard, Hart, David, Keller, James, and Darin Boville. 1997. "Investing in Innovation: Toward A Consensus Strategy for Federal Technology Policy." Paper for the Project on Technology Policy Assessment undertaken by the Science, Technology and Public Policy Program at Harvard University and sponsored by the Competitiveness Policy Council.

Bundesministerium für Umwelt, Naturschutz und Reaktorsicherheit, Germany, aluminum can and PET bottle recycling rates, <http://www. bmu.de/allgemein/aktuell/160.php> (October 7, 2003).

Busch, Andreas. 2001. Managing Innovation: Regulating the Banking Sector in a Rapidly Changing Environment. In *Success and Failure in Public Governance: A Comparative Analysis*. Edited by Mark Bovens, Paul Hart, and Peters B. Guy. Cheltenham: Edward Elgar, 311–325.

———. 2002. "National Filters: The Role of Institutions and Discourse in European Banking Regulation." Paper for the workshop Opening the Black Box: Europeanization, Discourse, and Policy Change at the University of Oxford Department of Politics and International Relations, November 23–24.

Callon, S. 1995. *Divided Sun: MITI and the Breakdown of Japanese High-Tech Industrial Policy, 1975–1993*. Stanford, CA: Stanford University Press.

Canton, E.J.F., Groot, H.L.F. de, and R. Nahuis. 1999. Vested Interest and Resistance to Technology Adoption. Discussion Paper No. 106 provided by Tilburg University, Center for Economic Research.

Casper, Steven, and Kettler Hannah, 2001. "National Institutional Frameworks and the Hybridization of Entrepreneurial Business Models: The German and UK Biotechnology Sectors." *Industry and Innovation*, Vol. 8: 5–13.

Castles, Franci, Gerritsen, Rolf, and Jack Vowles, eds. 1996. *The Great Experiment—Labour Parties and Public Policy Transformation in Australia and New Zealand*. NSW, Australia: Allen & Unwind Pty Ltd.

Central research Institute of Electric Power Industry, "Trend Analysis on Deregulation in the Electric Power Market" 2002, <http://criepi. denken.or.jp/eng/PR/Nenpo/1995E/seika95economics34E.html> (November 7, 2003).

Cerny, P.G. 1991. "The Limits of Deregulation: Transnational Interpenetration and Policy Change." *European Journal of Political Research*, Vol. 19, No. 2/3 (March/April): 173–196.

Chang, Ha-Joon, and Peter Evans. 2000. The Role of Institutions in Economic Change. Conference Paper at Venice, Italy, January 13–14.

Chapman, Robert M. 1998. *The Machine That Could: PNGV A Government Industry Partnership*. Washington, DC: RAND Publications.

Chilton, Kenneth. 1995. "Making Manufacturers Responsible for Recycling: Passing the Garbage Buck." Paper presented at the Solid Waste Association of North America 1995 U.S./Canadian Fed Solid Waste Management Conference, 17–32.

Clas-Otto, Wene. 2000. *Experience Curves for Energy Technology Policy*. Paris: OECD/IEA Publication.

Container Recycling Institute, various recycling rates, <http://www. container-recycling.org/> (October 7, 2003).

Cook, Paul, Kirkpatrick, Colin, Minogue, Martin, and Parker, David, eds. 2004. *Leading Issues in Competition, Regulation and Development*, Edward Elgar, Cheltenham, U.K.; Northampton, MA.

Crouch, Colin, and Wolfgang, Streeck 1997. "Introduction: The Future of Capitalist Diversity." In *Political Economy of Modern Capitalism: Mapping Convergence and Diversity*. Edited by Colin Crouch and Wolfgang Streeck. London: Sage, 1–18.

"Current Situation of Utilizing Photovoltaic Energy Use in Japan (Part 1)," *Global Change Info Japan*, <http://www.gcj.esto.or.jp/menu/ feature/9905.html> (March 20, 2003).

Danish Wind Industry Association (DWIA), *Danish Wind Energy in 2001— Statistics*, http://www.windpower.org/en/publ/index.htm (June 26, 2003).

Danish Wind Turbine Manufacturers Association (DWTMA), *The Wind Turbine Market in Denmark*, <http://www.windpower.org/en/ articles/wtmindk.htm> (August 25, 2003).

Department for Environment, Food and Rural Affairs. "Waste Strategy 2000," U.K., <http://www.defra.gov.uk/environment/waste/strategy/ cm4693/pdf/wastvol1.pdf> (March 23, 2003).

Department for Environment, Food and Rural Affairs, aluminum can and glass bottle recycling rates in various years, <http://www.defra.gov.uk/> (October 9, 2003).

Department of Trade and Industry, "The Renewable Obligation/NFFO" in 2002, <http://www.dti.gov.uk/> (August 15, 2003).

Doner, Richard F., and Ramsay Ansil. 2003. The Challenges of Economic Upgrading in Liberalising Thailand. In *States in the Global Economy: Bringing Domestic Institutions Back In*. Edited by Weiss Linda. New York: Cambridge University Press.

Dosi, Giovanni, Giannetti, Renato, and Pier Angelo Toninelli. 1992. *Technology and Enterprise in a Historical Perspective*. Oxford: Clarendon Press.

Doyle, Timothy, and Aynsley Kellow. 1995. *Environmental Politics and Policy Making in Australia*. Melbourne: Macmillan Education Australia Pty Ltd.

Duch, R.M. 1991. *Privatizing the Economy: Telecommunications Policy in a Comparative Perspective*. Ann Arbor: University of Michigan Press.

Dunn, William N., ed. 1986. *Policy Analysis: Perspectives, Concepts, and Methods*. Greenwich, CT: Jai Press Inc.

Dye, Thomas R. 1976. *Policy Analysis: What Governments Do, Why They Do It, and What Difference It Makes*. Tucaloosa: University of Alabama Press.

Ellis, James W. 1998. "Voting for Markets or Marketing for Voters: The Politics of Neo-liberal Economic Reform." Ph.D. Dissertation, Department of Government, Harvard University.

Ely, B. 1993. Savings and Loan Crisis. In *Fortune Encyclopedia of Economics*. Edited by David R. Henderson. New York: Warner Books, 369–375.

Energy Information Administration, "Country Analysis Briefs—Australia" in 2000, <http://www.eia.doe.gov/emeu/cabs/australi.html> (July 20, 2003).

Energy Research Centre of the Netherlands, costs between wind energy and conventional power, <http://www.ecn.nl/index.en.html> (September 12, 2002).

Energy South Australia. "Rebates and Grants," <http://www.sustainable.energy.sa.gov.au/pages/advisory/rebates/rebates.htm:sectID=106&tempID=58> (March 4, 2003).

Environmental Law Institute. 1997. *Research and Development Practices in the Environmental Technology Industry*. Washington, DC: Environmental Law Institute.

Environmental Management Centre. "Cleaner Technology Policy," <http://www.emcentre.com/> (February 8, 2003).

Environmental Protection Authority of New South Wales, various beverage container recycling rates, <http://www.epa.nsw.gov.au/> (October 12, 2003).

Environmental Protection Authority of South Australia, various beverage container recycling rates, <http://www.epa.sa.gov.au/> (October 12, 2003).

EUCC Coastal Guide. 2002. *Wind Power Special*, <http://www.coastal-guide.org/windpower/galambosi.html> (June 23, 2003).

European Aluminium Association, aluminum can recycling rates 1998, <http://www.eaa.net/> (October 15, 2003).

European Container Glass Federation, "News of Glass and Glass Recycling September 1997," Glass Gazette, <http://www.feve.org/index.php?option=content&task=view&id=4&Itemid=33> (October 15, 2003).

European Wind Energy Association, "The Report of Wind Energy in Europe—The Facts," < http://www.ewea.org/> (May 5, 2004).

Evans, Pete, Dietrich Reuschemeyer, and Theda Skocpol, eds. 1985. *Bringing the State Back In*. New York: Cambridge University Press.

Evans, Peter B. 1995. *Embedded Autonomy—States and Industrial Transformation*. Princeton, NJ: Princeton University Press.

——, ed. 1997. *State-Society Synergy: Government and Social Capital in Development*. University of California at Berkeley, International and Area Studies.

Fabbrini, S. 2003. "A Single Western State Model? Differential Development and Constrained Convergence of Public Authority Organization in Europe and America." *Comparative Political Studies*, Vol. 36: 653–678.

Fioretos, Karl Orfeo. 1998. Anchoring Adjustment: Globalization, Varieties of Capitalism, and the Domestic Sources of Multilateralism, Ph.D. Dissertation, Columbia University.

——, 2001. "The Domestic Sources of Multilateral Preference: Varieties of Capitalism in the European Community." In *Varieties of Capitalism*. Edited by P.A. Hall and D. Soskice. Oxford: Oxford University Press, 213–244.

Foreman-Peck, James, and Federico Giovanni, eds. 1999. *European Industrial Policy—The Twentieth-Century Experience*. Oxford: Oxford University Press.

Frieden, Jeffry. 1991. "Invested Interests: The Politics of National Economic Policies in a World of Global Finance." *International Organization*, Vol. 45, No. 4 (Fall): 425–451.

Friedman, Milton. 1981. *The Invisible Hand in Economics and Politics*. Singapore: Institute of Southeast Asian Studies.

Fukasaku, Yukiko. 2000. "Innovation for Environmental Sustainability: A Background." In *Innovation and the Environment: Sustainable Development*. Paris: OECD Publication, 24.

Gamble, A. 2000. "Economic Governance." In *Debating Governance: Authority, Steering, and Democracy*. Edited by J. Pierre. New York: Oxford University Press.

Garnant, Ross. 2002. "An Australia-United States Free Trade Agreement." *Australian Journal of International Affairs*, Vol. 56, No. 1: 123–141.

German Wind Energy Association, "Wind Energy 2003—Market Survey," <http://www.wind-energie.de/englischer-teil/english.htm> (June 8, 2004).

German Wind Energy Institute, "Statistics," <http://www.dewi.de/statistics.html> (March 25, 2004).

Gerston, Larry N., Cynthia Fraleigh, and Robert Schwab. 1988. *The Deregulated Society*. Pacific Grove, CA: Brooks/Cole Publishing Company.

Gesellschaft für Verpackungsmarktforschung mbH, Germany, aluminum can and PET bottle recycling rates in various years, <http://www.gvm-wiesbaden.de/> (October 15, 2003).

Gillham, Bill. 2000. *Case Study Research Methods*. London: Continuum.

Gilpin, Alan. 2000. *Environmental Economics, A Critical Review*. West Sussex, England: John Wiley & Sons Ltd.

Glass Packaging Institute, glass bottle recycling rates in various years, <http://www.gpi.org/> (October 15, 2003).

Glazer, Amihai, and Lawrence S. Rothenberg. 2001. *Why Government Succeeds and Why It Fails*. Cambridge, MA, London: Harvard University Press.

Godfrey, Bruce. 1993. "*Solar Energy Activities in Australia*," IEA SHC Solar Activities in IEA Countries Report, <http://www.iea-shc.org/publications/solar_energy_activities/australia.htm> (November 27, 2003).

Goodin, Robert E. 1996. *The Theory of Institutional Design*. New York: Cambridge University Press.

Green, K., and C. Morphet. 1977. *Research and Technology as Economic Activities*. London: Butterworths.

Grossman, Emiliano. 2003. "Bringing Politics Back in: The Role of Economic Interest groups in European Integration." *Cahiers Europeens De Sciences Po*, February.

Guellec, D., and B. van Pottelsberghe. 2000. The Impact of Public R&D Expenditure in Business R&D. STI Working Papers 2000/4. Paris: OECD Publication.

Hall, Peter A. 2003. "Aligning Ontology and Methodology in Comparative Research." In *Comparative Historical Analysis in the Social Sciences*. Edited by James Mahoney and Dietrich Rueschemeyer. Cambridge, U.K.; New York: Cambridge University Press.

———. 1986. *Governing the Economy: The Politics of State Intervention in Britain and France*. Cambridge, U.K.: Polity Press; in association with Oxford, U.K.: B. Blackwell.

Hall, Peter A., and Rosemary Taylor. 1996. "Political Science and the Three New Institutionalisms." *Political Studies*, Vol. 44: 936–957.

Hall, Peter A., and David Soskice, eds. 2001. *Varieties of Capitalism*. New York: Oxford University Press.

Hart, Jeffrey A. 1992. *Rival Capitalists*. New York: Cornell University Press.

Hatch, J. 1990. Container Deposit Legislation and the Control of Litter and Waste: Review of Information. Paper No. 14 of the Business Regulation Review Unit, Centre for South Australian Economic Studies, Adelaide.

Hay, Colin, and Daniel Wincott. 1998. "Structure, Agency and Historical Institutionalism." *Political Studies*, Vol. 46: 951–957.

Hayek, F.A. von. 1962. *In the Road to Serfdom*. London: Routledge & Kegan Paul.

Heaton, Jr., George R. 2000. "Workshop on Innovation and the Environment: Rapporteur's Report." In *Innovation and the Environment: Sustainable Development*. Pairs: OECD Publication, 9.

Hefeker, Carsten. 1997. Interest Groups and Monetary Integration. *The Political Economy of Exchange Regime Choice*. Boulder: Westview Press, 159.

Hills, Jill. 1986. *Deregulating Intercoms: Competition and Control in the United States, Japan and Britain*. Quorum, CT: Westport.

Hopgood, Stephen. 1998. *American Foreign Environmental Policy and the Power of the State*. New York: Oxford University Press.

House, Peter W. 1982. *The Art of Public Policy Analysis—The Arena of Regulations and Resources*. California: Sage Publications, Inc.

Huber, E., and J.D. Stephens. 1993. "Social Democracy, Christian Democracy, Constitutional Structure and the Welfare State." *American Journal of Sociology*, Vol. 99, No. 3: 711–749.

Hutchcroft, Paul D. 1998. *Booty Capitalis*. Ithaca and London: Cornell University Press.

Igora Genossenschaft. aluminum can and glass bottle recycling rates in various years, <http://www.igora.ch/> (October 16, 2003).

Ikcnbcrry, G. John, David A. Lake, and Michael Mastanduno, eds. 1988. *The State and American Foreign Economic Policy*. Ithaca: Cornell University Press.

Institute for Energy and Environmental Research 1999. "Wind Energy is far more Economical than Plutoniu," Press Release, http://www.ieer. org/reports/wind/wind-pr.html (January 13, 2002).

International Energy Agency. Energy Policies of IEA Countries 1998 Review. Paris: OECD/IEA Publication, 25–20.

International Energy Agency. 2000. "Key World Energy Statistics 2002," <http://www.iea.org/statist/index.htm> (July 18, 2003).

International Energy Agency. 2000. "Wind Energy Annual Report 2000," <http://www.iea.org/impagr/imporg/ann_reps/wind00/wind.htm> (January 15, 2003).

International Institute for Sustainable Development, "Wind Energy in Denmark," The Policy in Brief 2002, <http://iisd1.iisd.ca/greenbud/winden.htm> (March 8, 2003).

Isoard, S., and A. Soria. 2001. Technical Change Dynamics: Evidence from Emerging Renewable Energy Technologies. *Energy Economics*, Vol. 23, No. 6: 619–636.

Jaffe, Adam B., Richard G. Newell, and Robert N. Stavins. 1999. Energy-Efficient Technologies and Climate Change Policies: Issues and Evidence. Climate Issues Working Paper No. 19. Brandeis University—Resources for the Future and Harvard University—John F. Kennedy School of Government.

———. 2001. Technological Change and the Environment. Prepared as a Chapter of *The Handbook of Environmental Economics*. Working Paper, Harvard University, October.

John, Peter. 2002. Quantitative Methods. In *Theory and Methods in Political Science*. Edited by Marsh and Stoker. Hampshire and New York: Palgrave Macmillan.

Johnson, A., and S. Jacobsson. 2000, "The Diffusion of Renewable Energy Technology: An Analytical Framework and Key Issues for Research." *Energy Policy*, Vol. 28, No. 9: 625–640.

———. 2002. The Emergence of a Growth Industry: A Comparative Analysis of the German, Dutch and Swedish Wind Turbine Industries, DRUID Academy Winter 2002 Ph.D. Conference, Hotel Comwell Rebild Bakker Aalborg, Denmark, January 17–19.

Johnson, Chalmers. 1982. *MITI and the Japanese Miracle*. Stanford, CA: Stanford University Press.

———, ed. 1984. *The Industrial Policy Debate*. San Francisco, CA: Institute for Contemporary Studies.

Jordan, Grant, and Nigel Ashford, eds. 1993. *Public Policy and the Impact of the New Right*. London: Pinter Publishers.

Jordana, Jacint, and David Levi-Faur, eds. 2004. *The Politics of Regulation: Institutions and Regulatory Reforms for the Age of Governance*. Cheltenham; Northampton, MA: Edward Elgar.

———, special eds. 2005a. "The Rise of Regulatory Capitalism: The Global Diffusion of a New Order." *The Annals of the American Academy of Political and Social Science*, Vol. 598, March.

———. 2005b. "Preface: The Making of a New Regulatory Order." *The Annals of the American Academy of Political and Social Science*, Vol. 598, March 1–6.

Katzenstein, Peter J., ed. 1978. *Between Power and Plenty: Foreign Economic Policies of Advanced Industrial States*. Madison: University of Wisconsin Press.

Keil, Claudia. 1999. Environmental Policy in Germany. Resource Renewal Institute <http://www.rri.org/envatlas/europe/germany/de-index.html> (November 22, 2003).

Kemp, R. 2000. Technology and Environmental Policy: Innovation Effects of Past Policies and Suggestions for Improvement. In *Innovation and the Environment: Sustainable Development*. Paris: OECD Publication, 35–62.

———. 2002. Zero Emission Vehicle Mandate in California: Misguided Policy or Example of Enlightened Leadership? Paper submitted for publication in *Environment and Planning C: Government and Policy*.

Kemp, R., and A. Arundel. 1998. Survey Indicators for Environmental Innovation. IDEA paper series No. 8, STEP. Studies in Technology, Innovation and Economic Policy.

Kemp, R., J. Schot, and R. Hoogma. 1998. "Regime Shifts to Sustainability through Processes of Niche Formation: The Approach of Strategic Niche Management." *Technology Analysis and Strategic Management*, Vol. 10, No. 2: 175–195.

Kemp, R., B. Truffer, and S. Harms. 2000. Strategic Niche Management for Sustainable Mobility. In *Social Costs and Sustainable Mobility. Strategies and Experiences in Europe and the United States.* Edited by K. Rennings, O. Hohmeier, and R.L. Ottinger. Heidelberg: Physical Verlag, 167–187.

King, Desmond, and Stewart Wood. 1999. The Political Economy of Neoliberalism: Britain and the United States in the 1980s. In *Continuity and Change in Contemporary Capitalism.* Edited by Herbert Kitschelt, Peter Lange, Gary Marks, and John D. Stephen. Cambridge: Cambridge University Press, 371–397.

Kitschelt, Herbert. 1999b. *Continuity and Change in Contemporary Capitalism.* Cambridge: Cambridge University Press.

Krasner, Stephen D. 1978. United States Commercial and Monetary Policy: Unraveling the Paradox of External Strength and Internal Weakness. In Between Power and Plenty: Foreign Economic Policies of Advanced Industrial States. Edited by Peter Katzenstein. Madison: University of Wisconsin Press.

Krohn, Søren. 1998. "Creating a Local Wind Industry—Experience from Four European Countries," Danish Wind Turbine Manufacturers Association, <http://www.windpower.org/publ/soren.pdf> (July 22, 2003).

———. 2002. "Wind Energy Policy in Denmark: 25 Years of Success—What Now?" Danish Wind Industry Association, <http://www.windpower.org/media(493,1033)/Wind_energy_policy_in_Denmark:_25_years_of_success_what_now%3F.pdf> (July 22, 2003).

———. "The Wind Turbine Market in Denmark," Danish Wind Industry Association, <http://www.windpower.org/media(487,1033)/The_wind_turbine_market_in_Denmark.pdf> (July 22, 2003).

———. "The Great California Wind Rush," Danish Wind Industry Association, <http://www.windpower.org> (September 25, 2003).

Kuznets, S. 1960. Economic Growth of Small Nations. In *The Economic Consequences of the Size of Nations: Proceedings of a Conference Held by the International Economic Association.* Edited by E.A.G. Robinson. Toronto: MacMillan.

Laffer, Arthur B., Victor A. Canto, and Douglas H. Joines. 1983. *Foundations of Supply-Side Economics: Theory and Evidence.* New York: Academic Press.

Laundy, Philip. 1989. *Parliaments on the Modern World.* England: Dartmouth Publishing.

Lee, Steven Hugh. 1995. *Outposts of Empire: Korea, Vietnam and the origins of the Cold War in Asia—1949–1954.* Montréal: McGill-Queen's University Press.

Lehmann, Markus. 1999. "Private Institutions in Waste Management Policy and Their Antitrust Implications—The Case of Germany's Dual Management System," Paper from the *Preprints aus der Max-Planck-Projektgruppe Recht der Gemeinschaftsgüter Bonn.*

Lenehan, S. 1992. SA Container Deposit Legislation. Paper presented at the National Environmental Law Conference.

Levi-Faur, David. 2002. Comparative Research Designs in the Study of Regulation: How to Increase the Number of Cases without Compromising the Strengths of Case-Oriented Analysis. Paper presented at the International Workshop of the Centre on Regulation and Competition, September 4–6, 2002, University of Manchester.

———. 2004. Varieties of Regulatory Capitalism: Getting the Most out of the Comparative Method. Working Paper of the Research School of Social Sciences, Australian National University.

———. 2005. "The global Diffusion of Regulatory Capitalism." *Annals of the American Academy of Political and Social Sciences*, Vol. 598: 12–32.

Levin-Waldman, Oren M. 1996. *Reconceiving Liberalism—Dilemmas of Contemporary Liberal Public Policy*. University Press of Pittsburgh.

Lida, Tetsunari. 2000. "Strategies for Green Electricity in Japan." Working Paper of the Dept. of Environmental & Energy Policy Studies, Japan Research Institute, <http://www.fu-berlin.de/ffu/aktuelles/Salzburg/Iida.PDF> (February 18, 2003).

Lieberson, Stanley. 1985. *Making it Count: The Improvement of Social Research and Theory*. Berkeley: University of California Press.

Lijphart, Arend. 1975. "The Comparative-Cases Strategy in Comparative Research." *Comparative Political Studies*, Vol. 8: 158–177.

Lowndes, Vivien. 2002. "Institutionalism." In *Theory and Methods in Political Science*. Edited by Marsh and Stoker. New York: Palgrave Macmillan.

MacDougall, Terry, and Ikuo Kabashima. 1999. Japan: Democracy with Growth and Equity. In *Driven by Growth: Political Change in the Asia-Pacific Region*. Edited by James W. Morley. New York: M.E. Sharpe, 275–309.

Madlener, Reinhard, and Jochem Eberhard. 2002. "Regulatory and Institutional Innovation for the Promotion of Renewable Energy Use." Paper presented at the ENER Forum 3, Successfully Promoting Renewable Energy Sources in Europe. Budapest, Hungary, 6–7 June.

McCormick, Gavan, and Yoshio Sugimoto, eds. 1986. *Democracy in Contemporary Japan*. Armonk NY: M.E. Sharpe.

McKenzie, Richard, and Dwight Lee. 1991. *Quicksilver Capital*. New York: Free Press.

McLean, Paul D. 1996. "Patronage and Political Culture: Strategies of Self-Presentation." Dissertation, Department of Political Science, University of Chicago.

Mezey, Michael L. 1991. Parliament and Public Policy: An Assessment. In *Legislatures in the Policy Process—The Dilemmas of Economic Policy*. Edited by David, Olsen, and Michael Mezey. Cambridge: Cambridge University Press.

Milner, Helen. 1998. *Interests, Institutions and Information*. Princeton, NJ: Princeton University Press.

Mises, Ludwing von. 1963. *Human Action, A Treatise on Economics*. San Francisco: Fox & Wilkes, c1996. Also available online at <http://www.mises.org/humanaction.asp> (February 17, 2006).

Mitchell, Austin, ed. 1991. "Introduction: The View from the Lobbied: A Consumer's Guide." In *The Commercial Lobbyists*. Edited by Jordan Grant. Aberdeen University Press, 4.

Montpetit, Éric. 2003. *Misplaced Distrust: Policy Networks and the Environment in France, the United States, and Canada*. Vancouver and Toronto: UBC Press.

Moon, J. and C. Sharman. 2003. Western Australia. In *Australian Politics and Government: The Commonwealth, the States and the Territories*. Edited by Moon J. and C. Sharman. Melbourne: Cambridge University Press.

Moran, Michael. 1991. *The Politics of the Financial Services Revolution*. New York: St. Martins Press.

———. 1992. Regulatory Change in German Financial Markets. In *The Politics of German Regulation*. Edited by K. Dyson. Aldershot: Dartmouth Publishing.

———. 2001a. "Not Steering but Drowning: Policy Catastrophes and the Regulatory State." *Political Quarterly*, Vol. 72: 414–427.

———. 2001b. "The Rise of the Regulatory State in Britain." *Parliamentary Affairs*, Vol. 54: 19–34.

———. 2002. "Review Article: Understanding the Regulatory State." *British Journal of Political Science*, Vol. 32: 391–413.

Mukoyama, Toshihiko. 2002. A Theory of Technology Diffusion, Working Paper of Department of Economics, University of Rochester, January.

Nagel, Stuart S., ed. 1999. *The Substance of Public policy*. New York: Nova Science Publishers, Inc.

Naisbitt, J., and P. Aburdene. 1990. *Megatrends 2000: Ten New Directions for the 1990's*. New York: William Morrow and Company, Inc.

Nambu, Tsuruhiko. 1986. The Role of Government in the High Technology Industries. Australia-Japan Research Centre, Pacific Trade and Development Conference Special Paper.

Nash, John F. 1996c. *Essays on Game Theory*. Cheltenham, U.K.; Brookfield, VT: E. Elgar.

Nelson, C., and J. Gonzalez. 2001. *Improving California's Beverage Container Recycling Rate: A Case for Structural Program Changes*. Washington, DC: National Association of Schools of Public Affairs and Administration.

Nester, William R. 1997. *American Industrial Policy*. New York: St. Martin's Press.

———. 1990. *The Foundation of Japanese Power: Continuities, Changes, Challenges*. London: The Macmillan Press Ltd.

Nichols, Dick. 1999. *Environment, Capitalism & Socialism*. Australia: Resistance Books.

Nikkei Industry News. 1994. "MITI announces a list of light auto parts to be standardized, Tsusansho, kei buhin kyotsuka risuto kohyo." *Nikkei Sangyo Shimbun*, June 1: 11.

Niskanen, W. 1993. "Fiscal Effects on U.S. Economic Growth." In *Economic Policy, Financial Markets, and Economic Growth*. Edited by Benjamin Zycher and Lewis Solomon. Boulder: Westview, 235–246.

Noll, Roger G. 1989. "Economic Perspectives on the Politics of Regulation." In *Handbook of Industrial Organization*. Edited by Richard Schmalensee, and Robert Willig. Amsterdam: North-Holland, 1253–1287.

Nordhaus, William D. 1995. "The Ghosts of Climates Past and the Specters of Climate Change Future." *Energy Policy*, Vol. 23, No. 4/5: 269.

——. 1997. *The Swedish Nuclear Dilemma: Energy and the Environment*, Washington, DC: Resources for the Future.

——. 2000. *Modeling Induced Innovation in Climate-Change Policy*. Mimeo, Yale University.

North D.C. 1990. *Institutions, Institutional Change and Economic Performance*. Cambridge: Cambridge University Press.

O'Brien, A. 2001. Burning Forest for "Sustainable Energy." Are Policymakers Insane? The University of Sydney Union Campus Environment Committee. *Environmental Handbook*, Vol. 20.

OECD. 1996, Collection and Recycling Process of Waste Disposal, Contribution from Germany, Competition Policy Roundtables, OCDE/GD(96)22, No. 1, <http://www1.oecd.org/daf/clp/Roundtables/ ENVR04.HTM> (September 30, 2003).

——. 1997. *Diffusing Technology to Industry: Government Policies and Programmes*. Paris: OECD Publication.

——. 2000a. *Experience Curves for Energy Technology Policy 2000*. Paris: OECD Publication.

——. 2000b. *Innovation and the Environment—Sustainable Development 2000*. Paris: OECD Publication.

——. 2001a. *Local Partnerships for Better Governance*. Paris: OECD Publication.

——. 2001b. *Environmental Outlook*. Paris: OECD Publication.

——. 2002. *Governance for Sustainable Development—Five OECD Case Studies 2002*. Paris: OECD Publication.

Office of Transportation Technologies, Department of Energy. "PNGV Reports 1994–2001," <http://www.ott.doe.gov/oaat/pngv.html> (May 22, 2003).

Ogawa, Ken-Ichiro. "Japan PV Technology Status and Prospects," New Energy and Industrial Technology Development Organization, <http://www. euronet.nl/users/oke/PVPS/ar00/japan.htm> (April 5, 2003).

Ohmae, K. 1990. *The Borderless World*. London: Collins.

O'Neill, Kate. 2000. *Waste Trading among Rich Nations*. Massachusetts: The MIT Press.

Ormerod, P. 1999. "What Economics is Not." In *The Icon Critical Dictionary of Global Economics*. Edited by R. Beynon. Cambridge, U.K.: Icon Books Ltd.

Painter, Martin, and Shiu-Fai Wong. 2005. The Telecommunications Regulatory Regimes in Hong Kong and Singapore: When Direct State Intervention Meets Indirect Policy Instruments. Paper delivered to the Fourth International Convention of Asian Scholars, Shanghai Academy of Social Sciences, August 20–24.

Panagariya, Arvind, Paul R. Portney, and Robert M. Schwab. 1999. *Environmental and Public Economics*. Northampton, MA: Edward Elgar.

Peltzman, S. 1998. The Economic Theory of Regulation after a Decade of Deregulation. In *Reprinted in A Reader on Regulation*. Edited by R. Baldwin, C. Scott, and C. Hood. Oxford: Oxford University Press.

Pempel, T.J. 1977. "Introduction to" *Policymaking in Contemporary Japan*. Edited by T.J. Pempel. Ithaca: Cornell University Press, 13–21.

——. 1998. *Regime Shift: Comparative Dynamics of the Japanese Political Economy*. Ithaca, NY: Cornell University Press.

Pempel, T.J. and Michio Muramatsu. 1995. "The Japanese Bureaucracy and Economic Development: Structuring a Proactive Civil Service." In *The Japanese Civil Service and Economic Development: Catalysts of Change*. Edited by Kim, Hyung-Ki, Muramatsu, Michio, and T.J. Pempel. New York: Oxford University Press.

Pierson, Paul. 2000. Increasing Returns, Path Dependence and the Study of Politics. *American Political Science Review*, Vol. 94: 251–267.

Plastics and Chemicals Industries Association, "Plastics Recycling Survey 2001," <http://www.pacia.org.au/> (October 16, 2003).

Polt, Wolfgang. 2001. "The Role of Government in Networking." In *Innovative Networks, Co-operation in National Innovation Systems*. Edited by Schibany Andreas and Wolfgang Polt. Paris: OECD Publication.

Polverini, Melissa. 1998. "International PNGV-Equivalent Programs: Where does the United States stand?" *Journal of Engineering and Public Policy*, Vol. 2.

Porter, Gareth, and Brown Janet Welsh. 1996. *Global Environmental Politics*. Colorado: Westview Press, Inc.

Porter, M.E., and C. Van der Linde. 1995. "Toward a New Conception of the Environment-Competitiveness Relationship," *Journal of Economic Perspectives*, Vol. 9, No. 4 (Fall).

Powell, Walter W., and Paul J. DiMaggio, eds. 1991. The New Institutionalism in Organizational Analysis. Chicago: University of Chicago Press.

Przeworki, Adam, and Henry Teune. 1970. *The Logic of Comparative Social Inquiry*. New York: Wiley.

Quade, E.S., 3rd edition. 1989. *Analysis for Public Decisions*. New York: The RAND Corporation.

Qvortrup, Mads. 2002. *A Comparative Study of Referendums: Government by the People*. Manchester and New York: Manchester University Press.

Ragin, C.C. 2000. *Fuzzy-Set Social Science*. Chicago: Chicago University Press.

Rothstein, B. 1996. Political Institutions: An Overview. A New Handbook of Political Science. H.D.K.R.E. Goodin. Oxford: Oxford University Press, 133–166.

Ruggie, John Gerard, 1999. What Makes the World Hang Together? Neo-Utilitarianism and the Constructivist Challenge. In *Exploration and Contestation in the Study of World Politics*. Edited by Peter J. Katzenstein, Robert O. Keohane, and Stephen D. Krasner. Cambridge, MA and London, U.K.: MIT Press.

Saalfeld, Thomas. 1990. The West German Bundestag after 40 Years: The Role of Parliament in a "Party Democracy." In *Parliaments in Western Europe*. Edited by Philip Norton. London and Oregon: Frank Cass.

——. 1998. The German Bundestag: Influence and Accountability in a Complex Environment. In *Parliaments and Governments in West Europe*. Edited by Philip Norton. London and Oregon: Frank Cass.

Sakae, Tanemura. "Description of New Sunshine Program (NSS)," IEA, <http://www.iea-shc.org/publications/solar_energy_activities/japan.htm> (November 22, 2003).

Sandmo, Agnar. 2000. *The Public Economics of the Environment*. New York: Oxford University Press.

Schaller, Michael. 1985. *The American Occupation of Japan: The Origins of the Cold War in Asia*. New York: Oxford University Press.

Schmitter, Philippe, and Gerhard Lehmbruch. 1979. *Trends towards Corporatist Intermediation*. London: Sage.

Schultze, C. 1983. "Industrial Policy: A Dissent." *Brookings Review*, Vol. 2, No. 1 (October): 3–12.

Scitovsky, T. 1960. "International Trade and Economic Integration as a Means of Overcoming the Disadvantage of a Small Nation." In *The Economic Consequences of the Size of Nations*. Edited by E.A.G. Robinson. Proceedings of a conference held by the International Economic Association, Toronto: MacMillan.

Scott, Colin. 2004. Regulation in the Age of Governance: The Rise of the Post-regulatory State. In *The Politics of Regulation: Institutions and Regulatory Reforms for the Age of Governance*. Edited by Jordana Jacint and David Levi-Faur. Cheltenham; Northampton, MA: Edward Elgar, 145–174.

Sharman, Campbell, and Moon Jeremy. 2003. "One System or Nine?" In *Australian Politics and Government: The Commonwealth, the States and the Territories*. Edited by Jeremy Moon and Sharman Campbell. Melbourne: Cambridge University Press.

Shonfield, Andrew. 1965. *Modern Capitalism: The Changing Balance of Public and Private Power*. New York: Oxford University Press.

Short, Walter. 2000. "U.S. Renewable Energy Policy and Analysis," *National Renewable Energy Laboratory*, <http://www.nrel.gov/china/pdfs/re_forum/us_re_policy_and_analysis.pdf> (April 25, 2003).

Snyder, Lorid, Nolan H. Miller, and Robert H. Stavins. 2003. "The Effects of Environmental Regulation on Technology Diffusion: The Case of Chlorine Manufacturing." Working Paper Series No. RWP03–014, John F. Kennedy School of Government.

Solar Energy Research Group, "Current News 2000," <http://www.iwr.de/solar/neu/neu00_e.html> (April 24, 2003).

Sperling, Daniel. 2000. "Rethinking Public-Private R&D Partnerships: Lessons from PNGV." Project paper partly funded by The European Conference of Ministers of Transport (OECD) and University of California Transportation Center.

State of the Environment Australia, aluminum can recycling rates 1998, <http://www.deh.gov.au/soe/> (October 18, 2003).

Stavins, Robert N. 2004. "Introduction to the Political Economy of Environmental Refulation." In *The Political Economy of Environmental Regulation* (Working paper for the Introduction). Edited by Robert Stavins. Cheltenham: Edward Elgar.

Story, Johnathan, and Walter Ingo. 1997. *Political Economy of Financial Integration in* Europe. Manchester: Manchester University Press, 337.

Swann, D. 1988. *The Retreat of the State: Deregulation and Privatization in the UK and US*. New York: Harvester Wheatsheaf.

Swiss Federal Statistical Office, aluminum can and glass bottle recycling rates in various years, <http://www.bfs.admin.ch/bfs/portal/en/index.html> (October 18, 2003).

Taizo, Y. 1984. "Government in Spiral Dilemma: Dynamic Policy Interventions vis-à-vis Japanese Auto Firms, c. 1900–1960." In *An Economic Analysis of Japanese Firms*. Edited by Aoki Masahiko. Amsterdam: North Holland.

Thelen, Kathleen. 1996. "The Changing Character of Industrial Relations in Contemporary Europe." Paper presented at the 2nd workshop on "The New Europe: Rethinking the Collective Response to Change," Harvard University.

Thurow, L. 1985. "The Case for Industrial Policies in America." In *Economic Policy and Development: New Perspectives*. Edited by T. Shoshido and R. Sato. Dover, MA: Auburn House.

Tilton, John E. 2003. *On Borrowed Time? Assessing the Threat of Mineral Depletion*. Washington, DC: Resources for the Future.

Tisdell, Clement Allen. 1979. Priority Assessment in Science and Technology Policy. Report to the Australian Science and Technology Council, June.

Toffler, A. 1990. *Powershift*. New York: Bantam Books.

Tomlinson, Jim. 1990. *Public Policy and the Economy since 1900*. Oxford: Clarendon Press.

Tsebelis, George, and Money Jeannette. 1997. *Bicameralism*. Cambridge: Cambridge University Press.

U.S. Census Bureau, various bottle recycling rates, <http://www.census.gov/> (October 21, 2003).

U.S. Environmental Protection Agency, Office of Solid Waste, various recycling rates, <http://www.epa.gov/epaoswer/osw/> (October 21, 2003).

United Nations Development Programme, World Bank, World Resources Institute 2000. *World Resources 2000–2001: People and Ecosystems: The Fraying Web of Life*, World Resources Institute, Washington, DC.

Van Evera, Stephen. 1997. *Guide to Methods for Students of Political Science*. Ithaca and London: Cornell University Press.

Van Tiggelen, John. 2004. "All Ill Wind Blowing," *Good Weekend* (a magazine by the *Sydney Morning Herald*), September 4, 24.

Vig, Norman J., and Michael E. Kraft. 1990. Environmental Policy in the 1990, *Congressional Quarterly Inc.*, Washington, DC.

Vitols, Sigurt. 1996. "German Industrial Policy: An Overview," Discussion Paper FSI 96–321, *Wissenschaftszentrum Berlin für Sozialforschung*, <http://skylla.wz-berlin.de/pdf/1996/i96-321.pdf> (March 2, 2004).

——. 1997. "German Industrial Policy: An Overview." *Industry and Innovation*, Vol. 4, No. 1: 15–36.

——. 2001. Varieties of Corporate Governance: Comparing Germany and the UK. In *Varieties of Capitalism: The Institutional Foundations of Comparative Advantage*. Edited by Peter A. Hall and David Soskice. New York: Oxford University Press, 337–360.

Vogel, Steven K. 1996. *Freer Markets, More Rules*. New York: Cornell University Press.

——. 2001. The Crisis of German and Japanese Capitalism Stalled on the Road to the Liberal Market Model? *Comparative Political Studies*, Vol. 34, No. 10 (December): 1103–1133.

Wallace, David. 1995. *Environmental Policy and Industrial Innovation*. London: Royal Institute of International Affairs.

Wade, Robert. 1990. *Governing the Market—Economic Theory and the Role of Government in East Asian Industrialization*. Princeton, New Jersey: Princeton University Press.

Watanabe, Chihiro. 2003. *The Interaction between Technology and National Development: The Case of Japan, 1955–1995*. Working paper for the Director on Technology International Institute for Applied Systems Analysis, Department of Industrial Engineering and Management, Tokyo Institute of Technology, <http://crm.hct.ac.ae/events/archive/tend/028wata.html> (November 15, 2003).

Weingast, B.R. 1996. Political Institutions: Rational Choice Perspectives. In *A New Handbook of Political Science*. Edited by R.E. Goodin and H.D. Klingemann. Oxford.

Weiss, Linda, and Hobson John. 1995. *States and Economic Comparative Historical Analysis*. Cambridge: Polity Press.

Weiss, Linda. 1998. *The Myth of the Powerless State*. Cambridge; Oxford, U.K.: Polity Press.

——, ed. 2003. *States in the Global Economy*. Cambridge: Cambridge University Press.

Weitzman, Martin L. 1997. Sustainability and Technical Progress, *Scandinavian Journal of Economics*, Vol. 99, No. 1: 1–13.

Weller, Patrick, and Fleming Jenny. 2003. The Commonwealth. In *Australian Politics and Governments: The Commonwealth, the States and the Territories*. Edited by Jeremy Moon and Campbell Sharman. Cambridge: Cambridge University Press.

Williams, S., and Bateman B.G. 1995. *Power Plays*. Washington, DC: Investor Responsibility Research Center, 256.

Wilson, S. 2002. *Plastic Bottle Recycling in the UK*. Report for the Waste and Resources Action Programme. The Old Academy 21 Horse Fair Banbury Oxon OX16 0AH.

Wolferen, Karel van. 1989. *The Enigma of Japanese Power*. New York: Alfred A. Knopf.

Wood, David M. 1991. The British House of Commons and Industry Policy. In *Legislatures in the Policy Process—The Dilemmas of Economic Policy*. Edited by David Olsen and Michael Mezey. Cambridge: Cambridge University Press.

Wood, Stewart. 1997. "Capitalist Constitutions: Supply-Side Reforms in Britain and West-Germany 1960–1990." Ph.D. Dissertation, Department of Government, Harvard University.

———. 2001. "Business and Labor Market Policy." In *Varieties of Capitalism: The Institutional Foundations of Comparative Advantage*. Edited by Peter A. Hall and David Soskice. New York: Oxford University Press, 247–274.

World Bank. 1997. *World Development Report 1997*. Washington, DC: World Bank.

World Resources Institute, World Bank, United Nations Development Programme. 2000. *World Resources 2000–2001: People and Ecosystems: The Fraying Web of Life*. Washington, DC: World Resources Institute.

Woronoff, Jon. 1988. *Politics: The Japanese Way*. Basingstoke: Macmillan.

Wriston, W. 1992. *The Twilight of Sovereignty*. New York: Charles Scribner's Son.

Yamamura, Kozo. 1998. Distributional equity and adaptive capabilities: The resilience of German and Japanese capitalism? Paper presented at the Conference on Germany and Japan in the 21st Century: Strengths Turning into Weaknesses, Berlin, June 24–26.

Yamamura, Kozo and Wolfgang Streeck, eds. 2003. *The End of Diversity? Prospects for Japanese and German Capitalism*. Cornell: Cornell University Press.

Yan, Jinfen. 1998. *Utilitarianism in Chinese Thought*. St-Hyacinthe, Quebec: World Heritage Press Inc.

Ziegler, Nicholas. 1989. *The State and Technological Advance: Political Efforts for Industrial Change in the Federal Republic of Germany, 1972–1986*. Ph.D. Paper, Political Science, Harvard, Cambridge, MA.

Zysman, John. 1983. *Governments, Markets, and Growth; Financial Systems and the Politics of Industrial Change*. Ithaca: Cornell University Press.

INDEX